AIRO Springer Series

Volume 10

The **AIRO Springer Series** focuses on the relevance of operations research (OR) in the scientific world and in real life applications.

The series publishes peer-reviewed only works, such as contributed volumes, lectures notes, and monographs in English language resulting from workshops, conferences, courses, schools, seminars, and research activities carried out by AIRO, Associazione Italiana di Ricerca Operativa – Optimization and Decision Sciences: http://www.airo.org/index.php/it/.

The books in the series will discuss recent results and analyze new trends focusing on the following areas: Optimization and Operation Research, including Continuous, Discrete and Network Optimization, and related industrial and territorial applications. Interdisciplinary contributions, showing a fruitful collaboration of scientists with researchers from other fields to address complex applications, are welcome.

The series is aimed at providing useful reference material to students, academic and industrial researchers at an international level.

Should an author wish to submit a manuscript, please note that this can be done by directly contacting the series Editorial Board, which is in charge of the peer-review process.

THE SERIES IS INDEXED IN SCOPUS

Roberto Aringhieri · Francesca Maggioni ·
Ettore Lanzarone · Melanie Reuter-Oppermann ·
Giovanni Righini · Maria Teresa Vespucci
Editors

Operations Research for Health Care in Red Zone

ORAHS 2022, Bergamo, Italy, July 17–22

AIRO
ASSOCIAZIONE ITALIANA DI RICERCA OPERATIVA
OPTIMIZATION AND DECISION SCIENCE

Springer

Editors
Roberto Aringhieri
Dipartimento di Informatica
Università degli Studi di Torino
Torino, Italy

Ettore Lanzarone
Dipartimento di Ingegneria Gestionale,
dell'Informazione e della Produzione
Università degli Studi di Bergamo
Dalmine, Bergamo, Italy

Giovanni Righini
Dipartimento di Informatica
Università degli Studi di Milano
Milano, Italy

Francesca Maggioni
Dipartimento di Ingegneria Gestionale,
dell'Informazione e della Produzione
Università degli Studi di Bergamo
Dalmine, Bergamo, Italy

Melanie Reuter-Oppermann
University of Twente
Enschede, The Netherlands

Maria Teresa Vespucci
Dipartimento di Ingegneria Gestionale,
dell'Informazione e della Produzione
Università degli Studi di Bergamo
Dalmine, Bergamo, Italy

ISSN 2523-7047 ISSN 2523-7055 (electronic)
AIRO Springer Series
ISBN 978-3-031-38536-0 ISBN 978-3-031-38537-7 (eBook)
https://doi.org/10.1007/978-3-031-38537-7

This Springer imprint is published by the registered company Springer Nature Switzerland AG
The registered company address is: Gewerbestrasse 11, 6330 Cham, Switzerland

Contents

Editors and Contributors

About the Editors

Roberto Aringhieri is an associate professor of Operations Research at the University of Turin, Italy. He holds a M.S. in Computer Science and a Ph.D. in Mathematics for Economic Decisions and Operations Research from the University of Pisa. His main research interest focuses on quantitative methods applied to healthcare management. He is the co-chair of the EURO Working Group on Operational Research Applied to Health Services. He also serves as an associate editor for Operations Research for Healthcare.

Francesca Maggioni is an associate professor of Operations Research at the University of Bergamo, Italy. She holds a M.S. in Mathematics from Catholic University with laude and a Ph.D. in Pure and Applied Mathematics from the University of Milano Bicocca. Her main research interests are on optimization under uncertainty and applications in transportation and logistics. She is the chair of the EURO Working Group on Stochastic Optimization (EWGSO), the co-chair of the Stochastic Programming section of the Italian Operations Research Society (AIRO), and the secretary of the Stochastic Programming Society (SPS). She also serves as an associate editor for the journals: EURO Journal on Computational Optimization, Computational Management Science, Networks and TOP, An Official Journal of the Spanish Society of Statistics and Operations Research.

Ettore Lanzarone is a tenure-track associate professor at University of Bergamo. He is also a research collaborator at the Institute for Applied Mathematics and Information Technology "E. Magenes" (IMATI) of the National Research Council of Italy (CNR), where he worked as a researcher from 2011 to 2020. He is a member of the Centre Interuniversitaire de Recherche sur les Reseaux d'Entreprise, la Logistique et le Transport (CIRRELT), Montréal and Quebec City, Canada. He obtained his Ph.D. in Bioengineering and his master's degree in biomedical engineering cum laude in 2008 and 2004, respectively, at the Politecnico di Milano. His research activities focus on bioengineering, optimization and operations research, and stochastic models.

Melanie Reuter-Oppermann is an Assistant Professor for Operations Research in Healthcare at the University of Twente. Previously, she was a postdoctoral researcher at the Technical University of Darmstadt, Germany. She holds a diploma in Economathematics from the Technical University Kaiserslautern, Germany, and a Ph.D. in Industrial Engineering and Management with a focus on operations research for emergency medical services from the Karlsruhe Institute of Technology, Germany. She is a co-chair of the EURO Working Group on Operational Research Applied to Health Services (ORAHS). In 2021, she was awarded the Julius von Haast Fellowship from the Royal Society of New Zealand. Her main research interest is in operations research, analytics, artificial intelligence, and design science for healthcare services and logistics.

Giovanni Righini is a full professor of operations research at the University of Milan, Italy. He holds a M.S. in Electronic Engineering and a Ph.D. in Computer Engineering from the Technical University of Milan. His main research interest is in mathematical programming algorithms for combinatorial optimization problems. He is a member of the healthcare section of the Italian Operations Research Society (AIRO).

Maria Teresa Vespucci is an associate professor of Operations Research at the University of Bergamo, Italy. She holds a Ph.D. in Numerical Optimization from the University of Hertfordshire, UK. Her main research interest is in mathematical models for operational and investment problems in energy and industry.

Contributors

Kassim Said Abasse Département de Management – Faculté des sciences de l'administration, Université Laval, Quebec, QC, Canada;
Centre de Recherche CHU de Québec, Université Laval, Quebec, QC, Canada;
Centre de Recherche en Gestion Des Services de Santé, Université Laval, Quebec, QC, Canada;
VITAM Centre de Recherche en Santé Durable, Université Laval, Quebec, QC, Canada;
Centre de Recherche du CISSS de Chaudiére-Appalaches, CISSS de Chaudiére-Appalaches, Quebec, QC, Canada

Roberto Aringhieri Department of Computer Science, Università degli Studi di Torino, Turin, Italy

Michele Barbato Department of Computer Science, Università Degli Studi di Milano, Milano, Italy

Frédéric Bergeron Université Laval, Quebec, QC, Canada

Graziella Bonetti Director of Clinical Pathology Laboratory, ASST-Valcamonica, Esine, Italy

Giuliana Carello Department of Electronics, Information and Bioengineering (DEIB), Politecnico di Milano, Milan, Italy

Cristiano Carlevaro Spindox Spa, Milan, Italy

Alberto Ceselli Department of Computer Science, Università Degli Studi di Milano, Milano, Italy

Giuseppe Confessore Consiglio Nazionale delle Ricerche, Area Territoriale di Ricerca di Bologna, Bologna, Italy

André Côté Département de Management – Faculté des sciences de l'administration, Université Laval, Quebec, QC, Canada;
Centre de Recherche en Gestion Des Services de Santé, Université Laval, Quebec, QC, Canada;
VITAM Centre de Recherche en Santé Durable, Université Laval, Quebec, QC, Canada;
Centre de Recherche du CISSS de Chaudiére-Appalaches, CISSS de Chaudiére-Appalaches, Quebec, QC, Canada

Gloria De Luca ACT Operations Research IT Srl, Lomazzo (CO), Italy

Martina Doneda Department of Electronics, Information and Bioengineering (DEIB), Politecnico di Milano, Milan, Italy

Daniel Faccini Department of Management, Information and Production Engineering, University of Bergamo, Bergamo, Italy

Jean-Baptiste Gartner Département de Management – Faculté des sciences de l'administration, Université Laval, Quebec, QC, Canada;
Centre de Recherche CHU de Québec, Université Laval, Quebec, QC, Canada;
Centre de Recherche en Gestion Des Services de Santé, Université Laval, Quebec, QC, Canada;
VITAM Centre de Recherche en Santé Durable, Université Laval, Quebec, QC, Canada;
Centre de Recherche du CISSS de Chaudiére-Appalaches, CISSS de Chaudiére-Appalaches, Quebec, QC, Canada

Federico Gheza Research Fellow in General Surgery and Staff Surgeon, University of Brescia, Brescia, Italy

Maria Teresa Godinho Polytechnic Institute of Beja, School of Technology and Management, Beja, Portugal;
CMAFcIO, University of Lisbon, Lisboa, Portugal

Matthew Haren Département d'opérations et systèmes de décision – Faculté des sciences de l'administration, Université Laval, Quebec, QC, Canada;
Département de Management – Faculté des sciences de l'administration, Université Laval, Quebec, QC, Canada;
Centre de Recherche CHU de Québec, Université Laval, Quebec, QC, Canada;
Centre de Recherche en Gestion Des Services de Santé, Université Laval, Quebec, QC, Canada;
VITAM Centre de Recherche en Santé Durable, Université Laval, Quebec, QC, Canada;
Département de Marketing – Faculté des sciences de l'administration, Université Lava, Quebec, QC, Canada;
Centre de Recherche du CISSS de Chaudiére-Appalaches, CISSS de Chaudiére-Appalaches, Quebec, QC, Canada;
Centre Nutrition, Santé Et Société (NUTRISS), INAF, Université Laval, Quebec, QC, Canada;
Université Laval, Quebec, QC, Canada;
Laboratoire Humanis, EM Strasbourg-Business School, Université de Strasbourg, Strasbourg, France;
Centre Interuniversitaire de Recherche Sur Les Réseaux d'entreprise, La Logistique et Le Transport (CIRRELT), Université Laval, Quebec, QC, Canada;
Groupe de Recherche en Écologie Buccale (GREB), Université Laval, Quebec, QC, Canada

Paolo Landa Département d'opérations et systèmes de décision – Faculté des sciences de l'administration, Université Laval, Quebec, QC, Canada;
Centre de Recherche CHU de Québec, Université Laval, Quebec, QC, Canada;
Centre Interuniversitaire de Recherche Sur Les Réseaux d'entreprise, La Logistique et Le Transport (CIRRELT), Université Laval, Quebec, QC, Canada;
Groupe de Recherche en Écologie Buccale (GREB), Université Laval, Quebec, QC, Canada

Ettore Lanzarone Department of Management, Information and Production Engineering (DIGIP), University of Bergamo, Dalmine (BG), Italy

Célia Lemaire Laboratoire Humanis, EM Strasbourg-Business School, Université de Strasbourg, Strasbourg, France

Maria João Lopes Iscte-IUL University Institute of Lisbon, Business School, Lisboa, Portugal;
CMAFcIO, University of Lisbon, Lisboa, Portugal

Francesca Maggioni Department of Management, Information and Production Engineering, University of Bergamo, Bergamo, Italy

Filippo Manelli Director of Emergency Unit, ASST-Bergamo Est, Seriate, Italy

Catherine Paquet Centre de Recherche CHU de Québec, Université Laval, Quebec, QC, Canada;
Département de Marketing – Faculté des sciences de l'administration, Université Lava, Quebec, QC, Canada;
Centre Nutrition, Santé Et Société (NUTRISS), INAF, Université Laval, Quebec, QC, Canada

Marco Premoli Department of Computer Science, Università Degli Studi di Milano, Milano, Italy

Julia Resch Department of Operations and Information Systems, University of Graz, Graz, Austria

Melanie Reuter-Oppermann Department of High-tech Business and Entrepreneurship, University of Twente, Enschede, The Netherlands

Giovanni Righini Department of Computer Science, Università degli Studi di Milano, Milan, Italy

Elena Tànfani Dipartimento di Economia, Università degli studi di Genova, Genoa, Italy

Maria Teresa Vespucci Department of Management, Information and Production Engineering, Università degli Studi di Bergamo, Bergamo, Italy

Operations Research in the Red Zone

Roberto Aringhieri, Ettore Lanzarone, Francesca Maggioni, Melanie Reuter-Oppermann, Giovanni Righini, and Maria Teresa Vespucci

Keywords Operational research · Health-care · Optimization · Decision science · Machine learning

The EURO Working Group on Operational Research Applied to Health Services (ORAHS[1]) provides a network for researchers involved in the application of systematic and quantitative analysis in the planning and management of the health services sector. ORAHS was founded in 1975, as part of a program for the development of special interest groups within the European branch (EURO[2]) of the International

[1] http://orahs.di.unito.it/.
[2] https://www.euro-online.org/web/pages/103/working-groups.

R. Aringhieri (✉)
Department of Computer Science, Università degli Studi di Torino, Turin, Italy
e-mail: roberto.aringhieri@unito.it

E. Lanzarone · F. Maggioni · M. T. Vespucci
Department of Management, Information and Production Engineering, Università degli Studi di Bergamo, Bergamo, Italy
e-mail: ettore.lanzarone@unibg.it

F. Maggioni
e-mail: francesca.maggioni@unibg.it

M. T. Vespucci
e-mail: maria-teresa.vespucci@unibg.it

M. Reuter-Oppermann
Department of High-tech Business and Entrepreneurship, University of Twente, Enschede, The Netherlands
e-mail: m.n.reuter-oppermann@utwente.nl

G. Righini
Department of Computer Science, Università degli Studi di Milano, Milan, Italy
e-mail: giovanni.righini@unimi.it

R. Aringhieri et al. (eds.), *Operations Research for Health Care in Red Zone*,
AIRO Springer Series 10, https://doi.org/10.1007/978-3-031-38537-7_1

Federation of Operational research Societies (IFORS). The ORAHS annual meeting is open to anyone with a quantitative background and to those who have interest in the subject area. At the moment, the group has hundreds of members from more than thirty countries, mainly in Europe but also from overseas (e.g., USA, Canada, Brazil, Australia). Membership is open to any person interested in applying systematic, quantitative analysis to planning and management problems in the health area.

When applied to health care, operational research provides a rich diversity both in addressed problems and in solution methods: it allows to tackle complex logistic problems, such as ambulance routing, operating rooms scheduling and optimal location of resources in humanitarian logistics, as well as decision problems that are closer to clinical practice, such as therapy optimization, clinical pathways and appointments scheduling for periodical treatments. A stream of research concerns home care services and their integration with hospital-based services. The main challenges are to model, analyze and optimize complex and heterogeneous health care systems, allocating limited available resources to provide the most suitable type and location of treatment to each patient. This also calls for the analysis of decision processes and the definition of suitable metrics and optimization criteria, reconciling efficiency with equity. Many contributions have been recently developed in response to the COVID-19 pandemic emergency. Several investigations at the borders of operational research with computer science and engineering are also significant.

This volume stems from the 48th annual meeting (ORAHS 2022) that was hosted at the University of Bergamo, Italy, from July 17th to 22nd, 2022, organized by Ettore Lanzarone and Giovanni Righini. Among the contributions presented at ORAHS 2022, the papers collected in this volume have been selected, after a formal blind peer review process, to reflect the rich variety of operational research methods and applications.

The paper "A comparison of fairness metrics for health care problems" by M. Doneda et al. addresses the maybe basic, but crucial task of deciding what should be optimized. Defining the objectives, potentially considering fairness besides efficiency, is a decision science step that must precede any effort in mathematical optimization and algorithm design.

In the same vein, the paper "An overview of benefits and limitations of the process model notation applied for modeling patient healthcare trajectory" by P. Landa et al. presents the results of an extensive investigation on the use of Business Process Model Notation to describe and analyze health care processes. This paper explores the borders between operational research and process engineering, paving the way for a much-needed interdisciplinary approach.

Inspired and stimulated by the COVID pandemic, the papers "Machine Learning based Classification Models for COVID-19 Patients" by F. Maggioni et al. and "Integrating decision support tools in the COD-19 platform" by M. Barbato et al. stay at the borders between operational research and computer science. The efficient and fair allocation of scarce resources puts pressure on decision-makers in normal everyday life. When an emergency occurs, this pressure is raised to levels that are unmanageable without the support of mathematical models, optimization algorithms and an intelligent use of the available digital data.

The more classical operational research papers “A semi-online ambulance routing and scheduling problem with complex patient-vehicle relations” by J. Resch and “Towards a unified framework for routing and scheduling planning in an integrated continuous care unit” by M. Godinho and M. Lopes are devoted to patient transportation and home care. This shows a significantly increased interest in the optimization of the level of service provided to patients at the expense of the emphasis traditionally put on minimizing costs.

These selected contributions show how rich and promising operational research is, when applied to health care, in both its declinations as “decision science” to help decision-makers framing their problems and as “mathematical optimization” in a hard science setting, to design efficient algorithms and compute valuable solutions.

A Comparison of Fairness Metrics for Health Care Problems

Martina Doneda, Ettore Lanzarone, and Giuliana Carello

Abstract Fairness and balancing-related metrics can be modelled in several ways, and no single definition of fairness is universally accepted. From the application point of view, fairness is a rather common objective in problems where a decision maker needs to allocate resources/workloads to agents. This is also a rather common requirement in the health care field. For example, fair assignment constraints can be found in personnel rostering, and in workload or resource assignment problems. In particular, our analysis is motivated by the Blood Donor Appointment Scheduling (BDAS) problem, in which it is required to assign a number of donors of the different blood types over the days to achieve an as-constant-as-possible production of each blood type, hence enforcing balancing. Solving these problems may not be trivial, because quantities may not be divisible by the number of agents, additional constraints must be considered (e.g., limited capacities), or there are perturbations such as stochastic demands. In this work, we compare several approaches to model fair assignments and discuss their relevance to common health care service management problems. We show that different ways of modelling balancing constraints in the same problem produce different results, and we test them on benchmark instances inspired by the BDAS problem to derive insights on their performance.

Keywords Balanced assignments · Fairness · Health care management · Strategic and operational planning

M. Doneda (✉) · G. Carello
Department of Electronics, Information and Bioengineering (DEIB), Politecnico di Milano, Milan, Italy
e-mail: martina.doneda@polimi.it

G. Carello
e-mail: giuliana.carello@polimi.it

E. Lanzarone
Department of Management, Information and Production Engineering (DIGIP), University of Bergamo, Dalmine (BG), Italy
e-mail: ettore.lanzarone@unibg.it

R. Aringhieri et al. (eds.), *Operations Research for Health Care in Red Zone*, AIRO Springer Series 10, https://doi.org/10.1007/978-3-031-38537-7_2

1 Introduction

Obtaining *fair* solutions is often desirable when solving optimization problems. *Fairness* in assignment problems has been widely analyzed in several fields. Several fairness criteria have been proposed, and no single principle is universally accepted. In addition, there are also axiomatic notions of fairness, among which proportional and max-min fairness [1]. However, the concept of fairness is non-trivial and deserves to be studied [2]. From the application point of view, fairness is a rather common objective that models several situations in which a certain number of resources (or workloads) must be *fairly* distributed over a set of agents, e.g., operators, equipment or time-slots [3]. This is especially true in the field of health care provision, in which fairness and decision-making accountability are key requirements.

A fair solution is trivially obtainable if the resources or the workloads can be perfectly divided among the agents, i.e., if they are continuous or, in the case they are discrete, if their number is a multiple of the number of agents. However, the solution becomes non-trivial if other constraints apply, e.g., resources/workloads are discrete and non-multiple of the number of agents, capacity is limited, agents capacity is already partially assigned, or the amount of resources is not fixed.

In all these cases, the fairness requirement can be modelled in several ways. Some of these formulations directly depend on the nature of the resources being allocated. For instance, Castro-Silva and Gourdin [4] applied a fractioning-based technique to solve the load-balanced bin-packing problem, which is however only applicable with continuous quantities.

On the contrary, several fields deal with discrete quantities, e.g., in health care problems a treatment of a patient cannot be partitioned. Additionally, in many cases resources or demand items belong to different types, and a fair distribution must take this feature into account. For example, Erbayrak et al. [5] addressed a bin-packing problem with load balancing, in which items of the same family should be kept together. Or, in the nurse rostering problem, a set of shifts of different types (e.g., night, morning or afternoon shifts, weekday or weekend shifts) must be allocated to a set of nurses with different skills. This allocation should be done in such a way that the number of shifts of each type assigned to each nurse is as similar as possible among nurses, while respecting all other problem constraints [6].

Although the main objective of the nurse rostering problem is commonly to cover all shifts at minimum cost [7], in recent years several works [8] explored the concept of fair solutions [2, 9]. This is related to a very human aspect of dealing with personnel: even if a decision-making tool provides an exceptionally good roster, it is not surprising that people whose daily life is affected by this rostering solution may be displeased by it. For example, a cause of displease is the assignment of more or fewer *undesirable* shifts [10].

Moreover, the issue is not limited to staff rostering problems, but also encompasses other problems in the health care sector, hence justifying the interest of the research community [11]. For example, balancing problems arise in home care [12–14], inpatient bed allocation [15], and blood donation [16].

This work was originally motivated by the Blood Donor Appointment Scheduling (BDAS) problem [16, 17], in which it is required to assign a number of donors of the eight blood types (combinations of group and Rhesus factor) over the days, to achieve an as-constant-as-possible production of each type [18–20]. While studying this problem, it became clear that different formulations of this requirement were possible, and we took interest in analyzing how different formulations would perform.

We present seven formulations of a balanced distribution problem, which is a generalization of the BDAS problem, where a set of items of different types needs to be fairly allocated to a set of agents and a flexibility is considered for the amount of items to be allocated. We then compare the solutions in terms of balancing and overall assigned items, discussing the impact of the formulation on the performance with a focus on the application to the health care field.

2 Model Formulations

The addressed fair balancing problem is stated as follows, inspired by the BDAS problem in [16].

Amounts of items of different types $b \in B$ must be assigned to a set of agents $t \in T$ (decision variables x_b^t). Moreover, a_b^t items of type b may be already assigned to agent t for each b and t. In addition to them, the request to assign extra n_b^t items of type b exactly to agent t may arise for each b and t. Therefore, the total number y_b^t of items of type b assigned to t is given by $x_b^t + a_b^t + n_b^t$. Furthermore, the total already assigned items a_b^t and items to be assigned x_b^t over t $(\sum_{t\in T} (x_b^t + a_b^t))$ must lie within a range for each b, defined by a target level d_b and a flexibility parameter ε. Each agent t has a nominal capacity c_t and may also use an extra capacity $p_t \leq \mu\, c_t$, which, however, costs δ per unit used. Each item occupies a capacity r, the same for all items, and accordingly the capacity R_t globally occupied by the already assigned items a_b^t is a parameter given by $R_t = r \sum_{b\in B} a_b^t$. All sets, parameters and decision variables are detailed in Table 1. The constraints common to all formulations are:

$$y_b^t = x_b^t + n_b^t + a_b^t \qquad \forall t \in T, b \in B \qquad (1)$$

$$\lceil (1-\varepsilon)\, d_b \rceil \leq \sum_{t\in T} \left(x_b^t + a_b^t\right) \leq \lfloor (1+\varepsilon)\, d_b \rfloor \qquad \forall b \in B \qquad (2)$$

$$r \sum_{b\in B} (x_b^t + n_b^t) + R_t \leq c_t + p_t \qquad \forall t \in T \qquad (3)$$

$$p_t \leq \mu\, c_t \qquad \forall t \in T \qquad (4)$$

$$x_t^b, y_t^b \in \mathbb{N} \qquad \forall t \in T, b \in B \qquad (5)$$

Constraints (1) compute the overall number of items y_b^t whose values must be balanced over t for each b. Constraints (2) define the range the total number over $t \in T$

of already assigned items a_b^t and items to be assigned x_b^t must lay in. Constraints (3) limit the available capacity of each t. Finally, constraints (4) limit the extra capacity.

The objectives are to balance the number of items assigned to agents for each type, and to minimize the additional capacity required. The formulations differ in the way they include fair assignments, while they all limit p_t in the same way.

1. *Spread minimization*: this formulation minimizes the difference between the highest and the lowest y_t^b over all agents t for each type b:

$$\min \left\{ \sum_{b \in B} (\nu_b - \omega_b) + \frac{\delta}{|T|} \sum_{t \in T} p_t \right\} \tag{6}$$

s.t.

$$\nu_b \geq y_b^t \qquad \forall t \in T, b \in B \tag{7}$$

$$\omega_b \leq y_b^t \qquad \forall t \in T, b \in B \tag{8}$$

2. *min max*: this formulation minimizes the highest y_t^b over all agents t for each type b:

$$\min \left\{ \sum_{b \in B} \nu_b + \frac{\delta}{|T|} \sum_{t \in T} p_t \right\} \tag{9}$$

s.t.

$$\nu_b \geq y_b^t \qquad \forall t \in T, b \in B \tag{7}$$

3. *max min*: this formulation maximizes the lowest y_t^b over all agents t for each type b:

$$\max \left\{ \sum_{b \in B} \omega_b - \frac{\delta}{|T|} \sum_{t \in T} p_t \right\} \tag{10}$$

s.t.

$$\omega_b \leq y_b^t \qquad \forall t \in T, b \in B \tag{8}$$

4. $\pm$ *deviations*: this formulation minimizes the positive (z_b^{t+}) and negative (z_b^{t-}) deviations from a reference value z_b over T for each type b:

$$\min \left\{ \sum_{b \in B, t \in T} (z_b^{t+} + z_b^{t-}) + \delta \sum_{t \in T} p_t \right\} \tag{11}$$

s.t.

$$y_b^t = z_b + z_b^{t+} - z_b^{t-} \qquad \forall t \in T, b \in B \tag{12}$$

$$z_b^{t+} \geq 0 \qquad \forall t \in T, b \in B \tag{13}$$

$$z_b^{t-} \geq 0 \qquad \forall t \in T, b \in B \tag{14}$$

Variables z_b represent an intermediate value between the maximum and the minimum y_b^t over t. This can be the mean value of y_b^t if the summations of z_b^{t+} and $+z_b^{t-}$ over t are equal, or simply an intermediate value that allows to separate the contributions of z_b^{t+} and $+z_b^{t-}$ and compute them.

5. $+$ *deviations*: similar to the previous formulation, but it minimizes only the deviations z_b^{t+} above z_b:

$$\min\left\{\sum_{b\in B,t\in T} z_b^{t+} + \delta \sum_{t\in T} p_t\right\} \tag{15}$$

s.t.

$$y_b^t = z_b + z_b^{t+} \qquad \forall t \in T, b \in B \tag{16}$$

$$z_b^{t+} \geq 0 \qquad \forall t \in T, b \in B \tag{13}$$

6. $-$ *deviations*: similar to the previous two formulations, but it minimizes only the deviations z_b^{t-} below z_b:

$$\min\left\{\sum_{b\in B,t\in T} z_b^{t-} + \delta \sum_{t\in T} p_t\right\} \tag{17}$$

s.t.

$$y_b^t - z_b - z_b^{t-} \qquad \forall t \in T, b \in B \tag{18}$$

$$z_b^{t-} \geq 0 \qquad \forall t \in T, b \in B \tag{14}$$

7. *Absolute deviations* (used in [16]): this formulation minimizes the absolute deviations z_b^t from the average of y_t^b over T for each type b:

$$\min\left\{\sum_{b\in B,t\in T} z_b^{t} + \delta \sum_{t\in T} p_t\right\} \tag{19}$$

s.t.

$$- z_b^t\,|T| \leq \sum_{\tau\in T} y_b^\tau - y_b^t\,|T| \leq z_b^t\,|T| \qquad \forall t \in T, b \in B \tag{20}$$

To make the models comparable, the weight δ for the additional capacity $\sum_{t\in T} p_t$ is divided by $|T|$ when the first term of the objective function does not include the summation over $t \in T$ and is therefore expressed per agent.

Table 1 Sets, parameters and decision variables of the problem

Sets	
B	Types of items
T	Agents
Parameters	
d_b	Target for the total number over t of already assigned items a_b^t and items to be assigned x_b^t
ε	Flexibility parameter $\in [0, 1]$
a_b^t	Elements of type b already assigned to agent t
n_b^t	Additional assignments of items of type b exactly to agent t
r	Weight of each item
R_t	Overall weight of assigned items a_b^t
c_t	Nominal capacity of agent t
μ	Ratio between maximum extra and nominal capacity of any agent t
δ	Cost of using an extra capacity unit above c_t
Decision variables	
x_b^t	Number of items of type b assigned to agent t
y_b^t	Overall number of items of type b assigned to agent t
p_t	Used extra capacity of agent t above c_t

2.1 Comparing the Formulations

The solutions of the different formulations can be compared considering a common metric for all of them, e.g., the standard deviations of y_b^t over t for each b expressed through their average over b (SD_y). Moreover, they can be compared in terms of the total assigned items y_b^t over b and t (TOT_y).

However, the value of TOT_y is not considered in the above formulations, which can lead to equivalent optima with different TOT_y values, which may also affect SD_y. Thus, we solve an additional optimization problem to select the solution with the highest value of TOT_y among the equivalent optima.

The objective function of the second optimization problem is as follows:

$$\max \left\{ \sum_{b \in B, t \in T} y_b^t \right\} \tag{21}$$

while the constrains include those of the corresponding formulation for the first problem and an additional constraints that force the solution to be at least as good as that already obtained in terms of balancing. This means, depending on the formulation:

1. Spread minimization:

$$\sum_{b \in B} (\nu_b - \omega_b) + \frac{\delta}{|T|} \sum_{t \in T} p_t \leq \widehat{OF} \tag{22}$$

2. min max:

$$\sum_{b \in B} \nu_b + \frac{\delta}{|T|} \sum_{t \in T} p_t \leq \widehat{OF} \tag{23}$$

3. max min:

$$\sum_{b \in B} \omega_b - \frac{\delta}{|T|} \sum_{t \in T} p_t \geq \widehat{OF} \tag{24}$$

4. ± deviations:

$$\sum_{b \in B, t \in T} \left(z_b^{t+} + z_b^{t-}\right) + \delta \sum_{t \in T} p_t \leq \widehat{OF} \tag{25}$$

5. + deviations:

$$\sum_{b \in B, t \in T} z_b^{t+} + \delta \sum_{t \in T} p_t \leq \widehat{OF} \tag{26}$$

6. − deviations:

$$\sum_{b \in B, t \in T} z_b^{t-} + \delta \sum_{t \in T} p_t \leq \widehat{OF} \tag{27}$$

7. Absolute deviations:

$$\sum_{b \in B, t \in T} z_b^{t} + \delta \sum_{t \in T} p_t \leq \widehat{OF} \tag{28}$$

where $\widehat{OF}$ denotes the corresponding objective function value found by the first problem.

Obviously, this additional optimization is useful only with $\varepsilon > 0$; otherwise, $TOT_y = \sum_{b \in B} d_b$ regardless of the formulation.

3 Experimental Plan

The formulations were tested under different configurations with $\mu = 0.3$ and $\delta = 0.05$. Then, based on the BDAS problem where capacity is a working time, we set $r = 20$ min, and $c_t = 1350$ min per each t. For the problem size, we set $|B|$ to either 1 or 8 (corresponding to the cases of single blood type and 8 types in the BDAS problem, respectively), and $|T|$ to either 7, 97 or 140. Then, d_b was set to $56\,|T|$ overall items over T with $|B| = 1$, and to 1/8 of the previous value with $|B| = 8$ for each b, to make the alternatives comparable. For the perturbing factors a_b^t and

n_b^t we considered 3 levels: no factors (N); low level of factors (L); high level of factors (H). Each parameter a_b^t or n_b^t was set to 0 in level N, randomly generated from an independent uniform distribution with support {0, ..., 8} in level L, and from an independent uniform distribution with support {0, ..., 31} in level H.

Models were implemented in OPL and solved with CPLEX 12.8, and experiments were run on a Microsoft Windows 11 machine with an Intel® Core™ i5-1135G7 CPU and 16 GB of installed RAM. A Time Limit (TL) of 3600 seconds and no memory limit were set for all formulations.

We compared the formulations in terms of the computational time and the percentage optimality gap when TL is reached. Moreover, we also considered TOT_y and SD_y reached by solving the second optimization problem; when a gap greater than 0 was obtained, the best bound of TOT_y is also reported.

3.1 Results with $|B| = 1$

This setting neglects item types and only pursues a fair agent load.

Results in Table 2 show that all instances but one (with formulation ± deviation, $|T| = 97$ and perturbation level H) were solved to optimality in a negligible amount of time. *Ceteris paribus*, the formulations are not equivalent. With perturbation level N, all formulations but the max min with $|T| = 7$ and $|T| = 97$ provide perfectly balanced solutions with null SD_y. With level L, the only formulations that do not provide a null SD_y are the max min with all $|T|$ and the ± deviation with $|T| = 7$; again, TOT_y differs among the cases. Finally, with level H, a perfectly balanced is not provided by the absolute deviation formulation with all $|T|$, the min max with $|T| = 7$ and $|T| = 140$, and the ± deviation with $|T| = 7$ and $|T| = 97$. Moreover, with all levels and most $|T|$, TOT_y differs greatly among the cases.

We can also observe a predictable trend: adding perturbation factors generally increases SD_y when different from zero. At the same time, a less predictable phenomenon occurs: when the amount of perturbation factors increases to level H, the agents are more loaded (higher TOT_y). The max min formulation provides greater TOT_y in all cases, albeit in a tie with other formulations in some cases. On the other hand, it is the only one that fails at providing perfect balancing (null SD_y) at level N. Notably, with level H, the absolute deviation formulation consistently provides the smallest TOT_y and the highest SD_y, being the worst performer.

3.2 Results with $|B| = 8$

As mentioned, this setting is motivated by the BDAS problem, in which balancing is to be achieved among the 8 major human blood types [16].

Results are reported in Table 3. For perturbation level N, all formulations but the max min one are solved with perfect balancing (null SD_y). Differently from

Table 2 Results with $|B| = 1$: computational time and percentage optimality gap for the balancing problem; TOT_y and SD_y obtained solving the second problem. The number in parenthesis refers to the integer approximation of the best bound of TOT_y

$\|T\|$ level		Spread min.	Min. max.	Max. min.	± dev.	+ dev.	− dev.	Abs dev.
7	Time [s]	0.045	0.05	0.047	0.052	0.184	0.277	0.088
N	Gap [%]	0	0	0	0	0	0	0
	TOT_y	399	371	403	399	399	399	399
	SD_y	0.00	0.00	1.51	0.00	0.00	0.00	0.00
97	Time [s]	0.049	0.053	0.053	0.057	0.293	0.298	0.116
N	Gap [%]	0	0	0	0	0	0	0
	TOT_y	5432	5044	5527	5432	5432	5432	5432
	SD_y	0.00	0.00	0.14	0.00	0.00	0.00	0.00
140	Time [s]	0.056	0.058	0.064	0.054	0.134	0.087	0.158
N	Gap [%]	0	0	0	0	0	0	0
	TOT_y	7980	7280	7980	7980	7980	7980	7980
	SD_y	0.00	0.00	0.00	0.00	0.00	0.00	0.00
7	Time [s]	0.106	0.057	0.096	0.298	0.077	0.098	0.066
L	Gap [%]	0	0	0	0	0	0	0
	TOT_y	427	399	432	432	427	427	427
	SD_y	0.00	0.00	1.89	5.19	0.00	0.00	0.00
97	Time [s]	0.141	0.088	0.14	0.053	0.109	0.125	0.096
L	Gap [%]	0	0	0	0	0	0	0
	TOT_y	5820	5432	5897	5820	5820	5820	5820
	SD_y	0.00	0.00	0.41	0.00	0.00	0.00	0.00
140	Time [s]	0.139	0.077	0.136	0.056	0.107	0.093	0.123
L	Gap [%]	0	0	0	0	0	0	0
	TOT_y	8540	7840	8546	8540	8540	8540	8540
	SD_y	0.00	0.00	0.20	0.00	0.00	0.00	0.00
7	Time [s]	0.073	0.058	0.075	0.282	0.04	0.184	0.118
H	Gap [%]	0	0	0	0	0	0	0
	TOT_y	483	481	518	481	483	483	420
	SD_y	0.00	0.49	0.00	1.89	0.00	0.00	4.04
97	Time [s]	0.084	0.089	0.076	TL	0.164	0.084	1.445
H	Gap [%]	0	0	0	5.32%	0	0	0
	TOT_y	6499	6499	6984	6517 (6563)	6499	6499	5487
	SD_y	0.00	0.00	0.00	0.39	0.00	0.00	4.15
140	Time [s]	0.086	0.09	0.074	3.353	0.11	0.105	0.128
H	Gap [%]	0	0	0	0	0	0	0
	TOT_y	9520	9509	10220	9520	9520	9520	7980
	SD_y	0.00	0.27	0.00	0.00	0.00	0.00	5.48

$|B| = 1$, max min is not always the formulations that yields the highest TOT_y, being outperformed with all $|T|$ in level H. In particular, the—deviation formulation is the top performer in level H, providing the highest TOT_y with all $|T|$ and the smallest values of SD_y with $|T| = 97$ and $|T| = 140$. In contrast to $|B| = 1$, larger TOT_y were obtained because the pre-occupation of agent capacity (a_b^t) is distributed over 8 item types. Interestingly, equally balanced solutions yield again different levels of capacity exploitation; in particular, TOT_y are smaller with the min max formulation and perturbation level H. Then, the same phenomenon found when $|B| = 1$ regarding the trend of TOT_y among perturbation levels is observed.

With perturbation levels N and L, all cases are solved to optimality. With H, three formulations cannot prove the optimality of the found solution within TL when $|T| = 97$ and 140. To try to face this issue, we included the following constraints to break symmetries:

$$p_t \geq p_{t+1} \qquad \forall t \in T \setminus \{|T| - 1\} \tag{29}$$

It represents a valid inequality in level N, whilst in the other levels this constraint is a heuristic leading to a sub-optimal solution. Results in Table 4 show that, with level H, constraints (29) increased TOT_y in all instances of the min max formulation, in the instances with $|T| = 7$ and $|T| = 97$ of the max min formulation, and in the instance with $|T| = 97$ of the—deviation formulations.

4 Conclusions

In this work, we have analyzed different approaches to model a fair assignment problem.

We observed that different formulations are not equivalent and lead to different solutions, both in terms of balancing and of total items assigned when flexibility is given to the amount to be assigned. We considered a generalization of [16] for numerical experiments, deriving a series of insights that may find applications in other problems; future research will be devoted to including real-life features in the analysis to apply it to specific health care management problems.

To conclude, balancing is a common but tricky metric, which can lead to different effects depending on the formulation adopted. This may have a relevant impact, especially in health care where the uncertainty associated with the parameters is high. Therefore, we suggest to consider different formulations when addressing a problem and to evaluate not only the fairness performance of the formulations but also their impact on other metrics of interest for the specific problem. One or the other formulation will be preferred based on the specific problem addressed. For example, the absolute deviation formulation was employed for the BDAS problem.

Table 3 Results with $|B| = 8$: computational time and percentage optimality gap for the balancing problem; TOT_y and SD_y obtained solving the second problem. The numbers in parenthesis refer to the integer approximation of the best bound of TOT_y, while the asterisk denotes that TOT_y is that of the balancing problem because an integer solution was not found solving the second one

$\|T\|$ level		Spread min.	Min. max.	Max. min.	± dev.	+ dev.	− dev.	Abs dev.
7	Time [s]	0.161	0.146	0.172	0.128	0.542	0.366	0.32
N	Gap [%]	0	0	0	0	0	0	0
	TOT_y	392	392	400	392	392	392	392
	SD_y	0.00	0.00	0.35	0.00	0.00	0.00	0.00
97	Time [s]	0.347	0.251	0.249	0.175	0.541	0.576	2.05
N	Gap [%]	0	0	0	0	0	0	0
	TOT_y	5432	5432	5520	5432	5432	5432	5432
	SD_y	0.00	0.00	1.10	0.00	0.00	0.00	0.00
140	Time [s]	0.497	0.304	0.397	0.238	0.777	0.604	3.901
N	Gap [%]	0	0	0	0	0	0	0
	TOT_y	7840	7840	7976	7840	7840	7840	7840
	SD_y	0.00	0.00	0.95	0.00	0.00	0.00	0.00
7	Time [s]	0.226	0.274	0.203	0.301	0.324	0.784	1.272
L	Gap [%]	0	0	0	0	0	0	0
	TOT_y	421	421	433	413	405	421	421
	SD_y	0.50	0.50	1.36	0.49	0.43	0.50	0.50
97	Time [s]	1.019	0.486	0.382	TL	1.931	1080.41	26.386
L	Gap [%]	0	0	0	68.90%	0	0	0
	TOT_y	5696	5440	5937	5440 (5494)	5440	5696	5440
	SD_y	0.47	0.10	2.45	0.17	0.17	0.99	0.10
140	Time [s]	0.249	0.343	0.534	1.672	0.656	3.227	69.961
L	Gap [%]	0	0	0	0	0	0	0
	TOT_y	7842	7936	8564	7842	7842	7936	7846
	SD_y	0.04	0.28	2.30	0.04	0.06	0.28	0.07
7	Time [s]	0.906	0.441	1.552	5.564	2.174	1.116	3.365
H	Gap [%]	0	0	0	0	0	0	0
	TOT_y	488	481	488	488	488	504	488
	SD_y	0.71	1.23	0.73	0.71	0.71	0.74	0.71
97	Time [s]	3.433	1.19	0.893	TL	TL	1.515	TL
H	Gap [%]	0	0	0	49.55%	3.34%	0	4.57%
	TOT_y	6905	6624	6850	6726 (6850)	6726 (6793)	6957	6905
	SD_y	0.30	1.15	1.12	1.44	1.40	0.18	0.30
140	Time [s]	6.707	1.843	4.274	TL	TL	1.732	TL
H	Gap [%]	0	0	0	103.83%	35.83%	0	3.46%
	TOT_y	9886	9513	9941	9719 (9900)	9719 (9816)	10074	9886* (10124)
	SD_y	0.38	1.19	1.36	0.74	1.53	0.08	0.38

Table 4 Results with $|B| = 8$ and symmetry breaking constraints: computational time and percentage optimality gap for the balancing problem; TOT_y and SD_y obtained solving the second problem. The numbers in parenthesis refer to the integer approximation of the best bound of TOT_y, while the asterisk denotes that TOT_y is that of the balancing problem because an integer solution was not found solving the second one

$\|T\|$ level		Spread min.	Min. max.	Max. min.	± dev.	+ dev.	− dev.	Abs dev.
7	Time [s]	1.323	1.208	0.959	2.077	3.285	0.586	0.267
H	Gap [%]	0	0	0	0	0	0	0
	TOT_y	488	483	490	488	488	503	488
	SD_y	0.71	1.17	0.69	0.71	0.71	0.67	0.71
97	Time [s]	9.048	3.018	1.751	TL	TL	1.671	TL
H	Gap [%]	0	0	0	58.95%	3.96%	0	4.57%
	TOT_y	6905	6692	6905	6629* (6842)	6726 (6793)	6968	6905
	SD_y	0.30	1.01	0.30	0.60	0.60	0.14	0.30
140	Time [s]	10.549	5.485	3.732	TL	TL	4.11	TL
H	Gap [%]	0	0	0	120.37%	34.48%	0	3.46%
	TOT_y	9886	9520	9886	9520* (9909)	9719 (9816)	10043	9886* (9905)
	SD_y	0.38	1.21	0.43	0.63	1.00	0.18	0.38

References

1. Bertsimas, D., Farias, V.F., Trichakis, N.: The price of fairness. Oper. Res. **59**(1), 17–31 (2011)
2. Olivier, P., Lodi, A., Pesant, G.: Measures of balance in combinatorial optimization. 4OR **20**(3), 391–415 (2022)
3. Martello, S., Pulleyblank, W.R., Toth, P., De Werra, D.: Balanced optimization problems. Oper. Res. Lett. **3**(5), 275–278 (1984)
4. Castro-Silva, D., Gourdin, E.: A study on load-balanced variants of the bin packing problem. Discret. Appl. Math. **264**, 4–14 (2019)
5. Erbayrak, S., Özkır, V., Yıldırım, U.M.: Multi-objective 3D bin packing problem with load balance and product family concerns. Comput. Indus. Eng. 159, 107518 (2021)
6. Cheang, B., Li, H., Lim, A., Rodrigues, B.: Nurse rostering problems—a bibliographic survey. Eur. J. Oper. Res. **151**(3), 447–460 (2003)
7. De Causmaecker, P., Vanden Berghe, G.: A categorisation of nurse rostering problems. J. Sched. **14**(1), 3–16 (2011)
8. Glampedakis, A.: Fairness in Nurse Rostering Problem. Doctoral dissertation, University of Portsmouth (2018)
9. Böðvarsdóttir, E.B., Smet, P., Vanden Berghe, G., Stidsen, T.J.: Achieving compromise solutions in nurse rostering by using automatically estimated acceptance thresholds. Eur. J. Oper. Res. **292**(3), 980–995 (2021)
10. Heikkinen, E., Nikkonen, M., Aavarinne, H.: 'A good person does not feel envy': envy in a nursing community. J. Adv. Nurs. **27**(5), 1069–1075 (1998)
11. Xie, X., Lawley, M.A.: Operations research in health care. Int. J. Prod. Res. **53**(24), 7173–7176 (2015)

12. Lanzarone, E., Matta, A., Sahin, E.: Operations management applied to home care services: the problem of assigning human resources to patients. IEEE Trans. Syst., Man, Cybern. Part A: Syst. Hum. **42**(6), 1346–1363 (2012)
13. Carello, G., Lanzarone, E., Mattia, S.: Trade-off between stakeholders' goals in the home care nurse-to-patient assignment problem. Oper. Res. Health Care **16**, 29–40 (2018)
14. Bonomi, V., Mansini, R., Zanotti, R.: Fairness in home healthcare: can patient-centered and nurse-centered measures concur to the same goals? IFAC-PapersOnLine **55**(10), 3136–3141 (2022)
15. Cochran, J.K., Bharti, A.: Stochastic bed balancing of an obstetrics hospital. Health Care Manag. Sci. **9**(1), 31–45 (2006)
16. Baş, S., Carello, G., Lanzarone, E., Yalçındağ, S.: An appointment scheduling framework to balance the production of blood units from donation. Eur. J. Oper. Res. **265**(3), 1124–1143 (2018)
17. Yalçındağ, S., Baş Güre, S., Carello, G., Lanzarone, E.: A stochastic risk-averse framework for blood donation appointment scheduling under uncertain donor arrivals. Health Care Manag. Sci. **23**(4), 535–555 (2020)
18. Stanger, S.H., Yates, N., Wilding, R., Cotton, S.: Blood inventory management: hospital best practice. Transfus. Med. Rev. **26**(2), 153–163 (2012)
19. Baş, S., Carello, G., Lanzarone, E., Ocak, Z., Yalçındağ, S.: Management of blood donation system: Literature review and research perspectives. Springer Proc. Math. Stat. **169**, 121–132 (2016)
20. Baş Güre, S., Carello, G., Lanzarone, E., Yalçındağ, S.: Unaddressed problems and research perspectives in scheduling blood collection from donors. Prod. Plan. Control. **29**(1), 84–90 (2018)

An Overview of Benefits and Limitations of the Process Model Notation Applied for Modeling Patient Healthcare Trajectory

Paolo Landa, Jean-Baptiste Gartner, Matthew Haren, Célia Lemaire, Kassim Said Abasse, Catherine Paquet, Frédéric Bergeron, Elena Tànfani, and André Côté

Abstract The use of the business process model notation (BPMN) in modeling health care trajectory has been applied in the last decades in health care organizations for several aims, such as increasing efficiency and efficacy, improving patient outcomes and reducing costs. In scientific literature, many systematic reviews have been published about this topic and all of them have been inconclusive regarding

P. Landa (✉) · M. Haren
Département d'opérations et systèmes de décision – Faculté des sciences de l'administration, Université Laval, Quebec, QC G1V 0A6, Canada
e-mail: paolo.landa@fsa.ulaval.ca

J.-B. Gartner · M. Haren · K. S. Abasse · A. Côté
Département de Management – Faculté des sciences de l'administration, Université Laval, Quebec, QC G1V 0A6, Canada

P. Landa · J.-B. Gartner · M. Haren · K. S. Abasse · C. Paquet
Centre de Recherche CHU de Québec, Université Laval, Quebec, QC, Canada

J.-B. Gartner · M. Haren · K. S. Abasse · A. Côté
Centre de Recherche en Gestion Des Services de Santé, Université Laval, Quebec, QC, Canada

VITAM Centre de Recherche en Santé Durable, Université Laval, Quebec, QC, Canada

M. Haren · C. Paquet
Département de Marketing – Faculté des sciences de l'administration, Université Lava, Quebec, QC, Canada

J.-B. Gartner · M. Haren · K. S. Abasse · A. Côté
Centre de Recherche du CISSS de Chaudiére-Appalaches, CISSS de Chaudiére-Appalaches, Quebec, QC, Canada

M. Haren · C. Paquet
Centre Nutrition, Santé Et Société (NUTRISS), INAF, Université Laval, Quebec, QC, Canada

M. Haren · F. Bergeron
Université Laval, Direction Des Services-Conseils, Quebec, QC, Canada

M. Haren · C. Lemaire
Laboratoire Humanis, EM Strasbourg-Business School, Université de Strasbourg, Strasbourg, France

R. Aringhieri et al. (eds.), *Operations Research for Health Care in Red Zone*, AIRO Springer Series 10, https://doi.org/10.1007/978-3-031-38537-7_3

the effectiveness of BPMN in modeling health care trajectory. The objective of this study is to describe the use of BPMN in health trajectory, proposing a review that will define the main advantages and limitations of the decision modeling, highlighting areas of improvement on its ability to support clinical activities and decision-making processes. We propose the preliminary results of a review of the literature reporting the use of the BPMN approach for optimizing healthcare trajectories. The findings support the use of BPMN as a feasible and useful methodology to design and optimize clinical processes, as well as to automate tasks. However, the complexity of the healthcare domain makes process modeling such as BPMN a challenging task, for this reason the use of additional notations is proposed as alternative solutions to address the complexity.

Keywords Clinical modeling care pathway · Process optimization · Business process modeling notation

1 Introduction

In the last decades, the health care organisations had to manage several challenges. In most of the developed countries the health care systems have seen an increase of multiple morbidities, an increase of elder population, together with an increase of the demand for services, in both quantity and quality to provide to the citizens [1]. Many of these challenges focused on the need for efficient and effective management, as are required to the management improvement on outcomes and outputs, while considering a limited availability of resources and cost reduction. The health expenditure and financing have increased substantially in developed countries, such as the United States (US) and Canada, and the United Kingdom. In 2019, Canada spent 11.5% of its gross domestic product on healthcare expenses, according to the Organisation for Economic Cooperation and Development [2].

The quality of care and its delivery represents a serious challenge in health care systems, with high costs in hospital settings and a large impact on health outcomes [3–5]. Since the release of BPMN numerous national and international organizations have made repeated calls to national health care systems for the development of a framework for advancing the quality of care, ensuring safety, effectiveness, efficiency, patient centeredness, timeliness, and equity [6]. To satisfy these requirements, health care systems must develop solutions that enhance both efficiency and efficacy

P. Landa · M. Haren
Centre Interuniversitaire de Recherche Sur Les Réseaux d'entreprise, La Logistique et Le Transport (CIRRELT), Université Laval, Quebec, QC, Canada

Groupe de Recherche en Écologie Buccale (GREB), Université Laval, Quebec, QC, Canada

E. Tànfani
Dipartimento di Economia, Università degli studi di Genova, Genoa, Italy

of improving healthcare organization and patient outcomes while restraining costs. Efforts to improve clinical and care pathways have shown several benefits in terms of reduced variation in care, improved accessibility, quality, sustainability, and cost effectiveness of care [7–10]. Over the last decades, published articles have shown that mapping healthcare trajectories allowed to decrease the variation of professional practices and to standardized care processes [3–5]. This provided several benefits in the improvement of important characteristics of health care services: the accessibility, fluidity, quality, performance, and sustainability of healthcare services [7–10].

Considering these needs, during the last decades, several tools have been created and developed in order to support process improvement through process mapping. The Business Process Modeling Notation (BPMN) is one of the most important tools that were developed and utilised in this context. It consists of representing processes as a network of activities and tasks [11]. This approach has been supported by the Object Management Group (OMG) since 2005 and it has been adopted as an international standard by the International Organization for Standardization since 2012. The structure of this study is the following: in Sect. 2 the definition and contextualisation of BPMN and health trajectory, together with a literature review are reported. In Sect. 3 we describe the methodology related to the review and the application of the BPMN to health trajectory, while in Sect. 4 are reported the preliminary results. Finally, in Sect. 5 we report the conclusion and the future research of this study.

2 Literature Review

2.1 An Overview of the Health Care Trajectory

There are several definitions of health care trajectory in the literature [7, 13] and several include the term “pathway”. In this study we include the definition of patient-centered care pathway provided in a study published recently [13], as a long-term and complex managerial intervention adopting a systemic approach, for a well-defined group of patients who journey across the entire continuum of care, from prevention and screening to recovery or palliative care. This intervention: (i) prioritizes the centricity of patients and caregivers by analyzing the patient experience through their needs and expectations; (ii) supports the roles of professional actors involved in the care pathway by developing adherence to the patient-centered care approach; (iii) integrates a process of care approach through the modeling and improvement of the care pathway by continuously integrating the latest knowledge and information to support clinical decision-making; (iv) embeds coordination structures, (v) adapts to the contexts of both the physical and social structures; (vi) is supported by information systems and data management; and (vii) promotes the development of a learning health system to support the care pathway [13]. As described above, the health care trajectory is composed of a complex set of activities and characteristics that are difficult to represent. In the last decades the need of standardizing processes in health organizations increased showing the necessity of a standard to adopt.

2.2 The Business Processing Model Notation

The Business Process Model and Notation (BPMN) is one of the most used standards for business process modeling and maintained by the Object Management Group (OMG). The Business Process Modeling Notation is a graphical support used for the definition of complex event patterns. The basics of the BPMN notation is based on the types of graphical objects that comprise the notation and how they work together as part of a Business Process Diagram.

BPMN defines a Business Process Diagram (BPD), which is based on a flowcharting technique tailored for creating graphical models of business process operations. A Business Process Model, then, is a network of graphical objects, which are activities (i.e., work) and the flow controls that define their order of performance [14]. The adoption of this formal graphical notation represents the standard used to define a valid process which assures consistency and provides the same meaning as the textual description of the process [15]. In addition, the use of a modeling language is useful for translating a diagram into some machine-readable standard language. The standard is in its second version, and has been updated to BPMN 2.0. in 2010 (current version: 2.0.2, released December 2013) [12].

2.3 Decision Model and Notation

The Decision Model and Notation (DMN) 1.0 [16] was published in September 2015, while DMN 1.1 version was made available in May 2016. Numerous tool developers [17], already incorporated DMN modeling in their software packages, making the standard available for industry applications. DMN two levels can be used in conjunction. First, there is the decision requirement level, represented by the decision requirement diagram (DRD), which depicts the requirements of decisions and the dependencies between elements involved in the decision model. Secondly, there is the decision logic level, which presents ways to specify the underlying decision logic. The DMN standard provides an expression language S-FEEL (Simple Friendly Enough Expression Language), as well as boxed expressions and decision tables for the notation of the decision logic.

An increased interest in modeling decisions is present in scientific work in the field of process management, as illustrated by the body of recent literature on Decision Model and Notation (DMN) [18–20]. DMN is a declarative decision language, and it does not provide a decision resolution mechanism, which is left to the invoking context. The same holds for the processing and storage of outputs and intermediate results. With this recently introduced OMG standard for modeling decisions, it has become possible to extract decisions from processes and to model them separately according to the *Separation of Concerns* paradigm, hence enhancing understandability, scalability, and maintainability of processes, as well as that of underlying decisions [21–23]. DMN consists of two levels. First, the decision requirement level takes the form of a Decision Requirement Diagram (DRD) that depicts decisions and sub-decisions, business knowledge models, input data, and knowledge sources. It is used to portray the requirements of decisions and the dependencies between the different constructs in the decision model. Second, the decision logic level is used to specify the underlying decision logic. It is usually represented in the form of decision tables. The standard also provides a formal expression language FEEL (Friendly Enough Expression Language) that allows the execution of decision tables on a decision engine, as well as boxed expressions and a metamodel and schema. Decision tables are considered the core concept of DMN, and they contain the necessary information to automate decision-making. The DRD is mainly used to get a high-level understanding of the structure of the problem domain but does not contain additional information.

2.4 Case Management Model Notation

The Case Management Model and Notation (CMMN) version 1.1 was created by the Object Management Group (OMG) and published in December 2016. It is a complementary notation to the Business Process Model and Notation (BPMN) [24] which focuses on control flow to describe business processes. We will say that CMMN is declarative in which you describe 'what' is allowed and disallowed in the process; versus BPMN that is imperative in which you describe 'how' to do the process. BPMN, CMMN, and the Decision Model and Notation (DMN) are the three OMG business modeling notations. The case is the main concept in CMMN, and it is similar to a process. A case contains a case file (i.e., case data container) and it is described by a case plan (i.e., a model or diagram).

Case management (CM) was introduced as a tool for knowledge workers [25]. Its focus is on supporting unpredictable, knowledge-intensive and weakly structured processes. In contrast to classic processes, a certain goal and providing possibilities to choose from is more important than the way to achieve the goal itself. In a recent study [26], two distinct approaches for case management called Adaptive Case Management (ACM) and Production Case Management (PCM) are mentioned. With ACM, knowledge workers are allowed to manipulate the case for planning as well as at run-time without constraints. PCM, however, distinguishes between design-time, when the possible elements are developed, and run-time, when the case worker selects tasks and the case evolves [26].

Since CMMN is a relatively new standard, a brief introduction is provided here to aid comprehension. However, apart from ad hoc and event subprocesses, an explanation of BPMN is not included because it has been well documented elsewhere and the first version was published back in 2004. The literature review forms the foundation of the study and by means of argumentative and deductive reasoning, the differences and similarities between BPMN and CMMN are determined. The standard CMMN is composed of tasks, stages, milestones, and event listeners. Milestones are the accomplishments of the execution of the case instance and are very important in the process of understanding the progress of a particular case instance. The tasks are the execution of the actual work and there are four types: non-blocking human tasks, blocking human tasks, case task, and process task. The stages are containers similar to subprocesses in other workflow or process notations and they manage the complexity of the model by decomposing it into smaller manageable sets. Finally, the event listeners represent events in the workflow.

3 Methodology

The proceeding we propose a literature review based on the preliminary results of a scoping review that is under development [27]. Scoping review represents one of the best methods to map the available evidence regarding a specific topic, in our case study is the benefits and limitations of BPMN in modeling patient healthcare trajectory [28, 29]. The review has been conducted following the methodological frameworks proposed in [30] and improved in [31] and further refined by the Joanna Briggs Institute (JBI). A larger detail on the methodology is reported in the protocol previously published [27].

The Preferred Reporting Items for Systematic Reviews and Meta-Analyses Extension for Scoping Reviews (PRISMA-ScR) has also been adopted to structure the reporting of the review [32, 33]. In Fig. 1 is reported the selection of the studies following the PRISMA-ScR framework.

4 Preliminary Results on Benefits and Limitations

4.1 Analysis of the Literature

Our search strategy identified a total of 626 studies and 40 [36–65] met our eligibility criteria. The kappa agreement for study inclusion based on screening of titles and abstracts was 0.96. The PRISMA flow diagram is presented in Fig. 1. The data extraction process related to the scoping review included three main dimensions to analyse: study description, study results, and BPMN applications. Each dimension was split into several sub-dimensions, such as country and year of the study, BPMN model extension, BPMN linking extension, study design, type of health trajectory, study settings, key variables, aims of the study, expected results, study findings, study outcomes, study limitations, objectives for using BPMN, benefits of using BPMN, limitations of using BPMN. However, herein we will provide the preliminary results

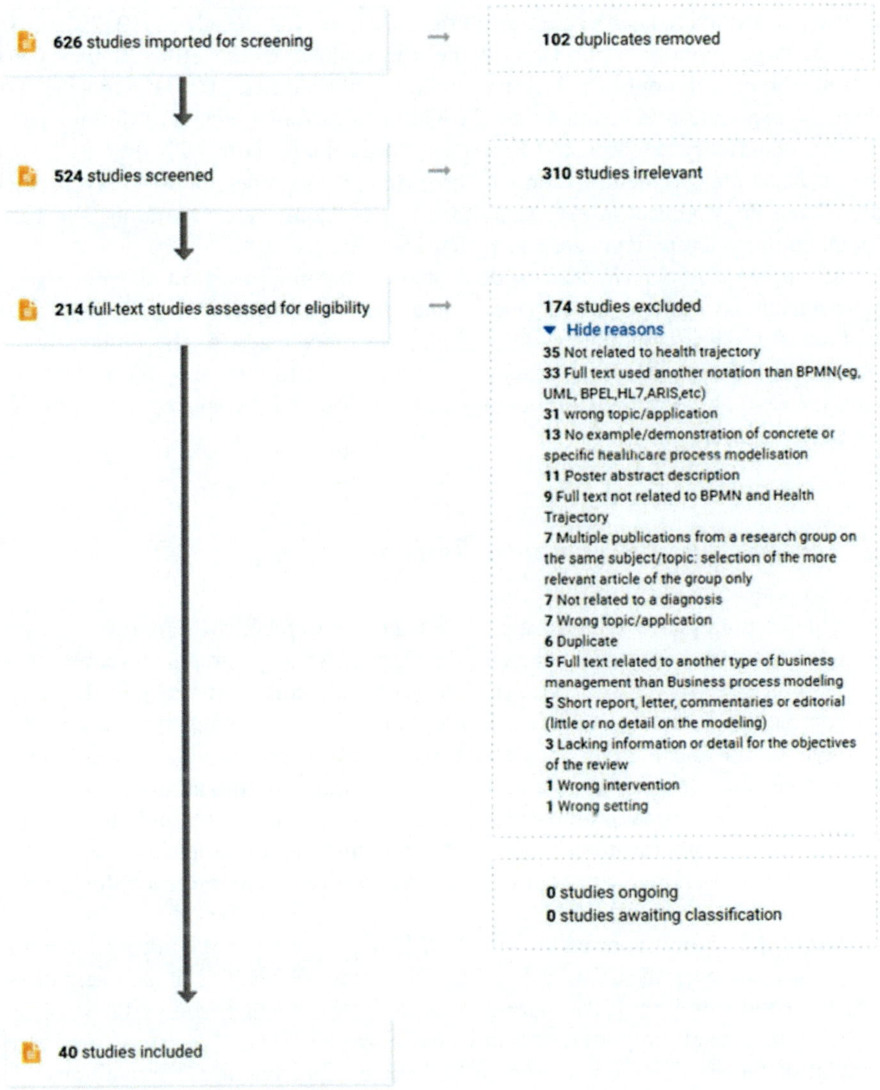

Fig. 1 Flow-chart for study selection

from the review, analysing some characteristics, such as the country of the study, the study design, the study aims, and the type of health trajectory.

Most of the studies were conducted in Europe (70%), while the remaining studies were conducted in South America (10%), North America (7.5%), Africa (7.5%) and Asia (5%). The study design was developed at conceptual level for 10 studies (25%), empirical for 9 studies (22.5%), experimental for 20 studies (50%), which only one was based on BPMN 2.0 and DMN 1.0 formalisms, and a theoretical (2.5%). The attention on BPMN to health care trajectories rose in the last decade (2010–2019), as it has the larger number of published studies (35 studies), while before 2010 we can find only three studies published and two studies published after 2020. The main areas where the application of health trajectory within BPMN approach is cardiology (six studies), respiratory diseases and Emergency department (both with four studies), while cancer care was identified only in three studies. The other areas were Neonatal care, Nephrology, Obstetrics, Pharmacy, Geriatrics, Home care, Operating Theatre, Ophthalmology, Outpatient care and Pediatrics.

The largest number of studies dedicated to the improvement of processes, improvement of system performances, and the improvement of pathways and resource use. The remaining studies (25%) were dedicated to the reduction of bottlenecks for overcrowding, design of hospital architecture, integrating knowledge, redesign and optimisation of surgical activities, and the conceptualisation of clinical pathways.

4.2 Discussion of Preliminary Results

Some of the studies have demonstrated the advantages of the BPMN approach, while in the others is not reported by the authors. The adoption of graphical representations of processes serves as an intuitive and more immediate reference for training and communication with healthcare professionals, or to clear represent processes that adhere and follow guidelines and evidence-based best practice. In addition, they support the automatization and the standardization of clinical procedures and decision-making, thereby promoting adherence to shared protocols and minimizing variability. These process models allow different types of process analyses and serve as a model for the automation of clinical and organizational activities and information flows.

Most of the studies reported the use of BPMN extensions or the integration with other tools. However, the use of extension shows some difficulties in the integration with other modelling tools, providing an issue in the development of the process representation, as also reported in a previous study [76].

Most of the studies consider the BPMN the best technique for modelling and representing business process, especially in health trajectory, where there is a large complexity in terms of resources and individuals involved in the care process. The BPMN was also considered easy to understand and suitable for both process experts and end users in health care settings. This enabled a clear comprehension to all of the active users and for their activities within the organization. The studies show that BPMN enables not just process information but in particular enable the communication of process information between participants in the process. In addition, BPMN visualises non-control-flow constructs such as trigger events, delays and messages as well as the structure of a business process (internal and collaborative levels).

Considering the largest part of the studies selected, BPMN is considered a cost-efficient, rational, standardised, intuitive, flexible tool. It can be modelled for any type of process structures, organizational structures and functional structures, enabling the tracking of process metrics for treatment and health trajectory patterns.

However, even if the studies have demonstrated that BPMN is a useful tool for modeling health care trajectories, it is clear that additional tools might improve the accessibility, fluidity, quality, performance and sustainability of health services. To face these challenges, several extensions have been developed and suggested supporting BPMN [34], including:

- Tangible Business Process Modeling (t.BPM) and BPMN + Clinical Pathway (BPMN4CP);
- BPMN + clinical guidelines for Clinical Decision Support (CDS);
- BPMN + Information and Communication Technology (ICT;
- BPMN + Surgical Intervention Extension (BPMN-six).

Despite the increase of those BPMN extensions, it is still difficult to find an extension that can fully address the demand of modeling healthcare trajectory. This explains the reason why the OMG has released new tools for case management and decision-making:

- CMMN (Case Management Model Notation), OMG release version 1.0 in March 2014 and version 1.1 in March 2016.
- DMN (Decision Model Nation), OMG release version 1.0 in September 2015 and June 2016 version 1.1
- BPM + Health™ (combination of BPMN + DMN + CMMN).

The BPM + Health is the one that provides the combination of BPMN with DMN and CMMN. The BPM + Health community of practice is an open ecosystem devoted to improving national and international health. It advances evidence-based medical practices and knowledge across the health sector by capturing and communicating best practices and clinical expertise. The BPMN + Health effectively makes it possible to respond to organizational and clinical concerns. In addition, it considers deterministic and non-deterministic events. This tool has the efficiency not only to map processes and activities, but it helps in the decision-making (given by the DMN) and helps in the description and management of health trajectory considering its complexity (given by the use of CMMN). As reported in [35], BPM + Health languages are supported by well-defined semantics and supporting documentation, providing models easy to read and understand.

The studies also presented a set of limitations for the use of BPMN. Few studies considered less useful for large studies with many patients, as the production of process maps is time-consuming. BPMN focuses on the existing activities and how they are performed, so it adapts only to previous defined activities in hospital. The method is not fully formalized, semantics are under-specified, models are not guaranteed to be interoperable between systems. The semantic of data objects remains unspecified and even left to the interpretation of the modelers and the behavior of data objects in BPMN models is decoupled from the control flow.

BPMN does not allow the modelling of weakly structured processes, which regularly occurs in the medical domain, and repeatable and discretionary time tasks are not possible to model in BPMN.

BPMN needs to be extended for the modeling of processes underlying medical practices, it provides little support to the expression of temporal properties and time constraints at a design level. BPMN does not specify how to model important and commonly used parameters of tasks and subprocesses such as time, risk, or performance measures. In addition, BPMN specification does not define multi-perspectivity, the possibilities of many views or layers of process models.

Considering the limitations herein described, we can state there is a great need for improvements for implementation and integration between BPMN and extensions, where was needed a feasible and flexible modelling tool, especially in health care trajectories where there is a large complexity. All the studies demand an increase of standards in the definition of BPMN, process metrics and semantics.

5 Conclusion and Future Research

BPMN is a well-defined standard that creates workflow diagrams to promote a common understanding and design of health care trajectories. Since its development, the application of BPMN has proven its efficiency in several areas, such as information technology, and supply chain management. However, its application in the health care organisation represents a challenge considering the complexity of the health care systems.

This study proposes to report an overview of benefits and limitations of BPMN applied for modeling patient healthcare trajectory. The results of this study come from an ongoing scoping review that is under development and will be completed. Considering the preliminary results, our findings highlight that the use of the BPMN can improve health care processes, resource use, and health care trajectories. However, even if there are several benefits, this standard presents some limitations that can be addressed using other tools recently provided by OMG.

Future research on this topic will be the finalisation of the scoping review that will provide a larger insight in the application of BPMN to health care trajectories. In addition, we want to explore if the BPM + Health can really address the ongoing challenges given by the BPMN and its extensions.

References

1. Organisation for Economic Co-operation and Development. Caring for quality in health . Organisation for Economic Co-operation and Development (2017). https://www.oecd.org/els/health-systems/Caring-for-Quality-in-Health-Final-report.pdf. Last accessed 09 Sept 2022

2. Organisation for Economic Co-operation and Development. Health at a glance (2021). https://www.oecd.org/health/health-at-a-glance/. Last accessed 09 Sept 2022
3. Gauld, R., Burgers, J., Dobrow, M., et al.: Healthcare system performance improvement: a comparison of key policies in seven high-income countries. J. Health Organ. Manag. **28**, 2–20 (2014)
4. Lee, S.E., Scott, L.D., Dahinten, V.S., et al.: Safety culture, patient safety, and quality of care outcomes: a literature review. West J. Nurs. Res. **41**, 279–304 (2019)
5. Kruk, M.E., Gage, A.D., Arsenault, C., et al.: High-Quality health systems in the sustainable development goals era: time for a revolution. Lancet Global Health **6**, e1196-1252 (2018)
6. Baker, A.: Crossing the quality chasm: a new health system for the 21st century. Br. Med. J. Publ. Group (2001)
7. Bergin, R.J., Whitfield, K., White, V., et al.: Optimal care pathways: a national policy to improve quality of cancer care and address inequalities in cancer outcomes. J. Cancer Policy **25**(100245) (2020)
8. Allen, D., Gillen, E., Rixson, L.: Systematic review of the effectiveness of integrated care pathways: what works, for whom, in which circumstances? Int. J. Evid. Based Healthc. **7**, 61–74 (2009)
9. Lodewijckx, C., Sermeus, W., Panella, M., et al.: Impact of care pathways for in-hospital management of COPD exacerbation: a systematic review. Int. J. Nurs. Stud. **48**, 1445–1456 (2011)
10. Alkandari, M., Ryan, K., Hollywood, A.: The experiences of people living with peripheral neuropathy in Kuwait—a process map of the patient journey. Pharmacy **7**(127) (2019)
11. Ferreira, A.S., Oliveira, G.R.: Business process modeling: a webibliominig perspective of architecture frameworks. Ind. J. Manag. Prod. **10**, 1159–1183 (2019)
12. Geiger, M., Harrer, S., Lenhard, J., Wirtz, G.: BPMN 2.0: The state of support and implementation. Futur. Gener. Comput. Syst. **80**, 250–262 (2018)
13. Gartner, J.B., Abasse, K.S., Bergeron, F., Landa, P., Lemaire, C., Côté, A.: Definition and conceptualization of the patient-centered care pathway, a proposed integrative framework for consensus: a Concept analysis and systematic review. BMC Health Serv. Res. **22**(558) (2022)
14. Rosing, M.V., White, S., Cummins, F.A., & Man, H.D.: Business Process Model and Notation—BPMN. The Complete Business Process Handbook, vol. I. Elsevier (2015)
15. Chinosi, M., Trombetta, A.: BPMN: An introduction to the standard. Comput. Standards Interfaces **34**(1), 124–134 (2012)
16. OMG: Decision Model and Notation 1.1 (2016). https://www.omg.org/spec/DMN/1.1/About-DMN/. Last accessed 09 Sept 2022
17. Eid-Sabbagh, R.H., Hewelt, M., Weske, M.: A tool for business process architecture analysis. In: Basu, S., Pautasso, C., Zhang, L., Fu, X. (eds.) Service-Oriented Computing. ICSOC 2013. Lecture Notes in Computer Science, vol. 8274. Springer, Berlin (2013)
18. Hasić, F., Devadder, L., Dochez, M., Hanot, J., De Smedt, J., Vanthienen, J.: Challenges in refactoring processes to include decision modelling. In: Teniente, E., Weidlich, M. (eds.) BPM 2017. LNBIP, vol. 308, pp. 529–541. Springer, Cham (2018)
19. De Smedt, J., Hasić, F., vanden Broucke, S.K.L.M., Vanthienen, J.: Towards a holistic discovery of decisions in process-aware information systems. In: Carmona, J., Engels, G., Kumar, A. (eds.) BPM 2017. LNCS, vol. 10445, pp. 183–199. Springer, Cham (2017)
20. Horita, F.E., de Albuquerque, J.P., Marchezini, V., Mendiondo, E.M.: Bridging the gap between decision-making and emerging big data sources: an application of a model-based framework to disaster management in brazil. Decis. Support Syst. **97**, 12–22 (2017)
21. Biard, T., Le Mauff, A., Bigand, M., Bourey, J.-P.: Separation of decision modeling from business process modeling using new decision model and notation (DMN) for automating operational decision-making. In: Camarinha-Matos, L.M., Bénaben, F., Picard, W. (eds.) PRO-VE 2015. IAICT, vol. 463, pp. 489–496. Springer, Cham (2015)
22. Hu, J., Aghakhani, G., Hasić, F., Serral, E.: An evaluation framework for design-time context-adaptation of process modelling languages. In: Poels, G., Gailly, F., Serral Asensio, E., Snoeck, M. (eds.) PoEM 2017. LNBIP, vol. 305, pp. 112–125. Springer, Cham (2017)

23. Hasić, F., De Smedt, J., Vanthienen, J.: Augmenting processes with decision intelligence: principles for integrated modelling. Decis. Support Syst. **107**, 1–12 (2018)
24. Marin, M.A, Lotriet, H., Van Der Poll, J.A.: Measuring Method Complexity of the Case Management Modeling and Notation (CMMN). In: Proceedings of the Southern African Institute for Computer Scientist and Information Technologists Annual Conference 2014 on SAICSIT 2014 Empowered by Technology (SAICSIT '14). Association for Computing Machinery, New York, NY, USA, pp. 209–216 (2014).
25. van der Aalst, W.M.P., Weske, M., Grünbauer, D.: Case handling: a new paradigm for business process support. Data Knowl. Eng. **53**(2), 129–162 (2005)
26. Swenson, K., Palmer, N. and Pucher, M., 2012. Case management: contrasting production versus adaptive. How knowledge workers get things done, pp.109–118
27. Kassim, S.A., Gartner, J.B., Labbé, L., et al.: Benefits and limitations of business process model notation in modelling patient healthcare trajectory: a scoping review protocol. BMJ Open **12**(060357) (2022)
28. Peterson, J., Pearce, P.F., Ferguson, L.A., Langford, C.A.: Understanding scoping reviews: Definition, purpose, and process. J. Am. Assoc. Nurse Pract. **29**, 12–16 (2017)
29. Munn, Z., Peters, M.D.J., Stern, C., et al.: Systematic review or scoping review? Guidance for authors when choosing between a systematic or scoping review approach. BMC Med. Res. Methodol. **18**(143) (2018)
30. Arksey, H., O'Malley, L.: Scoping studies: towards a methodological framework. Int. J. Soc. Res. Methodol. **8**, 19–32 (2005)
31. Levac, D., Colquhoun, H., O'Brien, K.K.: Scoping studies: advancing the methodology. Implement. Sci. **5**, 1–9 (2010)
32. Tricco, A.C., Lillie, E., Zarin, W.: A scoping review on the conduct and reporting of scoping reviews. BMC Med. Res. Methodol. **16**, 1–10 (2016)
33. Tricco, A.C., Lillie, E., Zarin, W., et al.: PRISMA extension for scoping reviews (PRISMA-ScR): checklist and explanation. Ann. Intern. Med. **169**, 467–473 (2018)
34. Grosskopf, A., Edelman, J., Weske, M.: Tangible business process modeling—methodology and experiment design. In: Rinderle-Ma, S., Sadiq, S., Leymann, F. (eds.) Business Process Management Workshops. BPM 2009. Lecture Notes in Business Information Processing, vol. 43. Springer, Berlin (2020)
35. Lario, R., Hasley, S., White, S,A,, Eilbeck, K., Soley, R., Huff, S., Kawamoto, K.: Utilization of BPM+ Health for the Representation of Clinical Knowledge: A Framework for the Expression and Assessment of Clinical Practice Guidelines (CPG) Utilizing Existing and Emerging Object Management Group (OMG) Standards. AMIA Annual Symposium Proceedings, pp. 687–696 (2021)
36. Ajmi, I., Zgaya, H., Gammoudi, L., Hammadi, S., Martinot, A., Beuscart, R., Renard, J.M.: Mapping patient path in the pediatric emergency department: A workflow model driven approach. J. Biomed. Inform. **54**, 315–328 (2015)
37. Ateetanan, P., Usanavasin, S., Shirahada, K., Supnithi, T.: From service design to enterprise architecture:the alignment of service blueprint and business architecture with business process model and notation. In: Hara, Y., Karagiannis, D. (eds.) Serviceology for Services. ICServ 2017. Lecture Notes in Computer Science, vol. 10371. Springer, Cham (2017). https://doi.org/10.1007/978-3-319-61240-9_19
38. Barbagallo, S., Corradi, L., de Ville de Goyet, J., Iannucci, M., Porro, I., Rosso, N., Tanfani, E., Testi, A.: Optimization and planning of operating theatre activities: an original definition of pathways and process modeling. BMC Med. Inform. Decis. Mak. **17**(15), 38 (2015). https://doi.org/10.1186/s12911-015-0161-7. PMID: 25982033; PMCID: PMC4436841.
39. Bazhenova, E., Zerbato, F., Oliboni, B., Weske, M.: From BPMN process models to DMN decision models. Inf. Syst. **83**, 69–88 (2019). ISSN 0306–4379, https://doi.org/10.1016/j.is.2019.02.001
40. Ben Hassen, M., Keskes, M., Turki, M., Gargouri, F.: BPMN4KM: design and implementation of a BPMN extension for modeling the knowledge perspective of sensitive business processes. Procedia Comput. Sci. **121**, 1119–1134 (2017). ISSN 1877–0509, https://doi.org/10.1016/j.procs.2017.12.121

41. Bisogno, S., Calabrese, A., Gastaldi, M. and Levialdi Ghiron, N.: Combining modelling and simulation approaches: How to measure performance of business processes. Bus. Process Manag. J. **22**(1), 56–74 (2016). https://doi.org/10.1108/BPMJ-02-2015-0021
42. Braun, R., Schlieter, H., Burwitz, M., Esswein, W.: BPMN4CP revised—extending BPMN for multi-perspective modeling of clinical pathways. In: 2016 49th Hawaii International Conference on System Sciences (HICSS), pp. 3249–3258 (2016). https://doi.org/10.1109/HICSS.2016.407
43. Braun, R., Schlieter, H., Burwitz, M., Esswein, W.: BPMN4CP: Design and implementation of a BPMN extension for clinical pathways. In: 2014 IEEE International Conference on Bioinformatics and Biomedicine (BIBM), pp 9–16 (2014). https://doi.org/10.1109/BIBM.2014.6999261
44. Çatal, N., Amyot, D., Michalowski, W., Kezadri-Hamiaz, M., Baslyman, M., Wilk, S., Giffen, R.: Supporting process execution by interdisciplinary healthcare teams: Middleware design for IBM BPM. Procedia Comput. Sci. **113**, 376–383 (2017). ISSN 1877–0509, https://doi.org/10.1016/j.procs.2017.08.350.
45. Combi, C., Oliboni, B., Zardiniy, A., Zerbato, F.: Seamless design of decision-intensive care pathways. In: 2016 IEEE International Conference on Healthcare Informatics (ICHI), pp 35–45 (2016). https://doi.org/10.1109/ICHI.2016.9
46. Combi, C., Oliboni, B., Zerbato, F.: Towards dynamic duration constraints for therapy and monitoring tasks. In: Ten Teije, A., Popow, C., Holmes, J., Sacchi, L. (eds.) Artificial Intelligence in Medicine. AIME 2017. Lecture Notes in Computer Science, vol. 10259. Springer, Cham (2017). https://doi.org/10.1007/978-3-319-59758-4_2
47. Crisan-Vida, M., Marcovici, A., Stoicu-Tivadar, L.: ICT solution for assisted diagnosis based on monitoring in cardiology departments. In: 2013 E-Health and Bioengineering Conference (EHB), pp. 1–4 (2013). https://doi.org/10.1109/EHB.2013.6707337
48. Cutting, E.M., Overby, C.L., Banchero, M., Pollin, T., Kelemen, M., Shuldiner, A.R., Beitelshees, A.L.: Using workflow modeling to identify areas to improve genetic test processes in the University of Maryland Translational Pharmacogenomics Project. In: AMIA Annual Symposium Proceedings, pp. 466–74. PMID: 26958179; PMCID: PMC4765659
49. de Bruin, J.S., Adlassnig, K.P., Leitich, H., Rappelsberger, A.: Separating business logic from medical knowledge in digital clinical workflows using business process model and notation and arden syntax. Stud. Health Technol. Inform. **248**, 17–24 (2018). PMID: 29726414
50. Eklund, P., Johansson, M., Karlsson, J., Åström, R.: BPMN and its semantics for information management in emergency care. In: 2009 Fourth International Conference on Computer Sciences and Convergence Information Technology, pp. 273–278. https://doi.org/10.1109/ICCIT.2009.27
51. Fox, J., Black, E., Chronakis, I., Dunlop, R., Patkar, V., South, M., Thomson, R.: From guidelines to careflows: modelling and supporting complex clinical processes. Stud. Health Technol. Inform. **139**, 44–62 (2008). PMID: 18806320
52. Haouari, G., Ayachi Ghannouchi, S.: Quality assessment of an emergency care process model based on static and dynamic metrics. Procedia Comput. Sci. **121**, 843–851 (2017). ISSN 1877–0509, https://doi.org/10.1016/j.procs.2017.11.109
53. Hewelt, M., Kunde, A., Weske, M., Meinel, C.: Recommendations for medical treatment processes: The PIGS approach. In: Fournier, F., Mendling, J. (eds.) Business Process Management Workshops. BPM 2014. Lecture Notes in Business Information Processing, vol. 202. Springer, Cham (2015). https://doi.org/10.1007/978-3-319-15895-2_2
54. Ilahi, L., Martinho, R., Ghannouchi, S.A., Domingos, D., Rijo, R.: Similarity based approach for comparing home healthcare processes models in Portugal. Procedia Comput. Sci. **100**, 1250–1259 (2016)
55. Jimenez-Molina, A., Gaete-Villegas, J., Fuentes, J.: ProFUSO: Business process and ontology-based framework to develop ubiquitous computing support systems for chronic patients' management. J. Biomed. Inform. **82**, 106–127 (2018). https://doi.org/10.1016/j.jbi.2018.04.001. Epub 2018 Apr 5. PMID: 29627462

56. Kopecky, M., Tomaskova, H.: The business process model and notation used for the representation of Alzheimer's Disease patients care process. Data **5**(1), 16 (2020). https://doi.org/10.3390/data5010016
57. Lamine, E., Tawil, A.R.H., Bastide, R., Pingaud, H.: Ontology-based workflow design for the coordination of homecare interventions. In: Camarinha-Matos, L.M., Afsarmanesh, H. (eds.) Collaborative Systems for Smart Networked Environments. PRO-VE 2014. IFIP Advances in Information and Communication Technology, vol. 434. Springer, Berlin (2014). https://doi.org/10.1007/978-3-662-44745-1_67
58. Nan, S., Van Gorp, P., Lu, X., et al.: A meta-model for computer executable dynamic clinical safety checklists. BMC Med. Inform. Decis. Mak. **17**, 170 (2017). https://doi.org/10.1186/s12911-017-0551-0
59. Neumann, J., Franke, S., Rockstroh, M., Kasparick, M., Neumuth, T.: Extending BPMN 2.0 for intraoperative workflow modeling with IEEE 11073 SDC for description and orchestration of interoperable, networked medical devices. Int. J. Comput. Assist. Radiol. Surg. **14**(8), 1403–1413 (2019). https://doi.org/10.1007/s11548-019-01982-6. Epub 2019 May 4. PMID: 31055764
60. Peres Penteado, A., Molina Cohrs, F., Diniz Hummel, A., Erbs, J., Maciel, R.F., Feijó Ortolani, C.L., de Aguiar, R.B., Torres, P.I.: Kidney transplantation process in Brazil represented in business process modeling notation. Transplant Proc. **47**(4), 963–966 (2015). https://doi.org/10.1016/j.transproceed.2015.03.044. PMID: 26036495
61. Poulymenopoulou, M., Papakonstantinou, D., Malamateniou, F., Vassilacopoulos, G.: Adaptive healthcare processes for personalized emergency clinical pathways. Stud. Health Technol. Inform. **205**, 423–427 (2014). PMID: 25160219
62. Ramos-Merino, M., Álvarez-Sabucedo, L.M., Santos-Gago, J.M., et al.: A BPMN based notation for the representation of workflows in hospital protocols. J. Med. Syst. **42**, 181 (2018). https://doi.org/10.1007/s10916-018-1034-2
63. Rodriguez-Loya, S., Aziz, A., Chatwin, C.: A service oriented approach for guidelines-based clinical decision support using BPMN. Stud Health Technol Inform. **205**, 43–47 (2014). PMID: 25160142
64. Ruiz-Fernández, D., Marcos-Jorquera, D., Gilart-Iglesias, V., Vives-Boix, V., Ramírez-Navarro, J.: Empowerment of patients with hypertension through BPM, IoT and remote sensing. Sensors (Basel) **17**(10), 2273 (2017). https://doi.org/10.3390/s17102273.PMID:28976940;PMCID:PMC5677452
65. Russo, V., Ciampi, M., Esposito, M.: A business process model for integrated home care. Procedia Comput. Sci. **63**, 300–307 (2015). ISSN 1877–0509, https://doi.org/10.1016/j.procs.2015.08.347
66. Scheuerlein, H., Rauchfuss, F., Dittmar, Y., Molle, R., Lehmann, T., Pienkos, N., Settmacher, U.: New methods for clinical pathways-Business Process Modeling Notation (BPMN) and Tangible Business Process Modeling (t.BPM). Langenbecks Arch. Surg. **397**(5), 755–761 (2012). https://doi.org/10.1007/s00423-012-0914-z. Epub 2012 Feb 24. PMID: 22362053
67. Sooter, L.J., Hasley, S., Lario, R., Rubin, K.S., Hasić, F.: Modeling a clinical pathway for contraception. Appl. Clin. Inform. **10**(5), 935–943 (2019). https://doi.org/10.1055/s-0039-3400749. Epub 2019 Dec 11. PMID: 31860113; PMCID: PMC6924335.
68. Svagård, I., Farshchian, B.A.: Using business process modelling to model integrated care processes: experiences from a european project. In: Distributed Computing, Artificial Intelligence, Bioinformatics, Soft Computing, and Ambient Assisted Living. IWANN 2009. Lecture Notes in Computer Science, vol. 5518. Springer, Berlin (2009). https://doi.org/10.1007/978-3-642-02481-8_140
69. Tomaskova, H., Kopecky, M., Maresova, P.: Process cost management of Alzheimer's disease. Processes **7**(9), 582 (2019). https://doi.org/10.3390/pr7090582
70. Vandborg, M.P., Edwards, K., Kragstrup, J., Vedsted, P., Hansen, D.G., Mogensen, O.: A new method for analyzing diagnostic delay in gynecological cancer. Int. J. Gynecol. Cancer. **22**(5), 712–717 (2012). https://doi.org/10.1097/IGC.0b013e31824c6d0e. PMID: 22453532

71. Vinci, A.L.T., Barbosa, F., de Pádua, S.I.D., Rijo, R., Alves, D.: The process of outpatient care of children and adolescents in a tertiary-level hospital specializing in pediatrics: A case study focused on identifying opportunities for improvement with the aid of modeling using BPMN. Knowl. Process Manag. **25**, 193–206 (2018). https://doi.org/10.1002/kpm.1571
72. Weber, P., Filho, J.B.F., Bordbar, B., Lee, M., Litchfield, I., Backman, R.: Automated conflict detection between medical care pathways. J. Softw. Evol. Proc. **30**, e1898 (2018). https://doi.org/10.1002/smr.1898
73. Wiemuth, M., Junger, D., Leitritz, M.A., et al.: Application fields for the new Object Management Group (OMG) Standards Case Management Model and Notation (CMMN) and Decision Management Notation (DMN) in the perioperative field. Int. J. CARS **12**, 1439–1449 (2017). https://doi.org/10.1007/s11548-017-1608-3
74. Zerbato, F., Oliboni, B., Combi, C., Campos, M., Juarez, J.M.: BPMN-based representation and comparison of clinical pathways for catheter-related bloodstream infections. Int. C. Healthcare Inf. **2015**, 346–355 (2015). https://doi.org/10.1109/ICHI.2015.49
75. Rodrigues, A.L., Torres, F.B.G., Santos, E.A.P., Cubas, M.R.: Process modeling: technological innovation to control the risk for perioperative positioning injury. Rev. Bras. Enferm. **74**(suppl 6), e20200145 (2021). English, Portuguese. https://doi.org/10.1590/0034-7167-2020-0145. PMID: 34495212
76. Zarour, K., Benmerzoug, D., Guermouche, N., Drira, K.: A systematic literature review on BPMN extensions. Bus. Process. Manag. J. **26**(6), 1473–1503 (2020). https://doi.org/10.1108/BPMJ-01-2019-0040

Machine Learning Based Classification Models for COVID-19 Patients

Francesca Maggioni, Daniel Faccini, Federico Gheza, Filippo Manelli, and Graziella Bonetti

Abstract The SARS-CoV-2 pandemic has pushed the National Health Service to extraordinary pressure, causing situations of imbalance between the request and availability of assistance. When the number of patients exceeds the available resources, doctors need to establish priorities among the patients to be treated. This paper describes novel data-driven optimization models to support doctors' decisions to solve one of the main problems encountered during the first months of the COVID-19 pandemic: predict the mortality risk for COVID-19 in order to address the most appropriate therapeutic path. The models are trained using clinical data obtained at the access to the Emergency Department of 150 SARS-CoV-2 infected patients admitted to ASST-Valcamonica (Brescia, Italy), in March 2020. To handle the uncertainty in data, we formulate robust and distributionally robust optimization models and compare their performance with other 31 different classification models from the literature, including decision trees, discriminant analysis, support vector machines, logistic regression, nearest neighbors, and naive Bayes. Numerical results show that robust formulations allow to achieve higher levels of accuracy with respect to the corresponding deterministic ones. The best prediction results are obtained with an optimized decision tree model, allowing to identify the most important factors. The tool can be used after triage to more accurately assess the severity of a COVID-19

F. Maggioni (✉) · D. Faccini
Department of Management, Information and Production Engineering, University of Bergamo, Bergamo, Italy
e-mail: Francesca.Maggioni@unibg.it

D. Faccini
e-mail: Daniel.Faccini@unibg.it

F. Gheza
Research Fellow in General Surgery and Staff Surgeon, University of Brescia, Brescia, Italy
e-mail: Federico.Gheza1@unibs.it

F. Manelli
Director of Emergency Unit, ASST-Bergamo Est, Seriate, Italy
e-mail: Filippo.Manelli@asst-bergamoest.it

G. Bonetti
Director of Clinical Pathology Laboratory, ASST-Valcamonica, Esine, Italy
e-mail: Graziella.Bonetti@asst-valcamonica.it

R. Aringhieri et al. (eds.), *Operations Research for Health Care in Red Zone*, AIRO Springer Series 10, https://doi.org/10.1007/978-3-031-38537-7_4

patient's condition, allowing doctors to optimize patient accommodation by identifying those in need of intensive care and those instead of sub-intensive care.

Keywords Decision support · COVID-19 application · Data analysis and risk management

1 Introduction

Coronaviruses are RNA viruses, known to cause disease in humans and animals, ranging from common cold to more severe and even deadly respiratory infections. Strains of betacoronavirus have been identified in 2003 and 2012 as causing *Severe Acute Respiratory Syndrome* (SARS) and also the *Middle East Respiratory Syndrome* (MERS). *Coronavirus Disease 2019* (COVID-19) is an infectious disease caused by *SARS Coronavirus 2* (SARS-CoV-2) according to the International Committee on Taxonomy of Viruses [4]. SARS-CoV-2 is a new type of coronavirus and its nucleic acid sequence is different from SARS-CoV and MERS-CoV. At April 6th 2020, the last statistics of the *World Health Organization* (WHO) revealed that COVID-19 had already affected over 554.550 people from almost every country worldwide, causing as many as 47.687 death, while as of May 31st 2022 cases reached 532.941.055 with around 6.313.941 deaths. As of June 2022, Italy is the 9th of the most deeply involved country, for the number of cases (17.421.410), deaths (166.697) and especially the mortality rate, 2.765 deaths every one million citizens [10].

The clinical manifestations of COVID-19 infection include fever, myalgia, dry cough, dyspnoea, fatigue and less frequently headache, diarrhoea, nausea, vomiting as well as anosmia and ageusia. In severe cases, COVID-19 can rapidly turn into acute respiratory distress syndrome, septic shock, bleeding, coagulation dysfunction, metabolic acidosis and death [6]. The role of laboratory medicine has always been of critical importance during viral outbreaks [7], due to its ability of identifying possible clinical predictors of progression towards severe and fatal forms of infections [1]. Some parameters which have already been shown to have an influence on COVID-19 patients outcomes are: aspartate aminotransferase, lactate dehydrogenase, C-reactive protein, neutrophils and lymphocyte counts, haemoglobin, platelets count, procalcitonin, high sensitive cardiac troponin I, urea, creatinine, cardiac biomarkers, and partial hromboplastin time [3]. The definition of the strongest predictors enables risk stratification among patients at high or low risk of mortality, allowing for improved clinical situational care.

Therefore, the aim of the present study is to analyze the common laboratory abnormalities in patients with COVID-19 employing the most recent *Machine Learning* (ML) techniques, in order to identify which are the parameters most likely to classify patients between those who are well and those who are unlikely to survive. From the ML standpoint, a great variety of algorithms have been devised to address the classification problem: *Decision Trees* (DT), *Discriminant Analysis* (D), *Logistic Regression* (LR), *Naive Bayes* (NB), *Support Vector Machines* (SVM), and *k-Nearest*

Neighbors (k-NN) classifiers, etc. An underlying assumption of ML approaches is to handle noise in data only indirectly (or implicitly) at the moment of classifying. This assumption, however, is not always practical as real-world observations are often plagued by noise (e.g., due to limited precision of collecting instruments, measurement mistakes in data gathering, sampling errors, etc.) and two of the main paradigms to deal with problems affected with uncertainty are given by *Robust Optimization* (RO) [2] and *Distributionally Robust Optimization* (DRO) [9].

In this paper, therefore, to predict COVID-19 mortality risk and support doctors' decisions of addressing the most appropriate patients' therapeutic path, we perform a computational comparison of literature ML classification models, including novel robust and distributionally robust SVM formulations from [5] that explicitly handle data uncertainty in the training set.

The paper is organized as follows. Section 2 presents robust and distributionally robust optimization models for SVM; Sect. 3 describes data collection and reports the experimental study on prediction of COVID-19 mortality risk, while conclusions are provided in Sect. 4.

2 Methods

In this section we aim to build a ML model to support doctors' decisions of "stratifying patients' clinical risk", hence to predict the mortality risk of COVID-19 patients so as to guide the best diagnostic and therapeutic care path. Specifically, we will first recall the deterministic formulation from [8]. To handle uncertainty in data features due to limited precision of collecting instruments or measurement mistakes, we further consider the robust and distributionally robust counterparts proposed in [5].

2.1 Deterministic Formulation

Let $X := \{x^{(1)}, x^{(2)}, \dots, x^{(I)}\} \subseteq \mathbb{R}^n$ and $Y := \{y^{(1)}, y^{(2)}, \dots, y^{(J)}\} \subseteq \mathbb{R}^n$ be two sets which correspond, respectively, to surviving COVID-19 patients (class 0) and to COVID-19 patients who died within their hospital stay (class 1). For every patient n features based on clinical and laboratory data are available. The hyperplane $a^\top x = \gamma$ (with $a \in \mathbb{R}^n$, $\gamma \in \mathbb{R}$ later denoted with (a, γ)) that separates sets X and Y is found solving the following deterministic SVM optimization problem (see [8]):

$$\begin{aligned}
\min_{a,\gamma,z_X,z_Y} \quad & \|a\|_1 + \nu(e^\top z_X + e^\top z_Y) & \\
\text{s.t.} \quad & a^\top x^{(i)} \leq \gamma - 1 + z_{x^{(i)}} & i = 1, \dots, I \\
& a^\top y^{(j)} \geq \gamma + 1 - z_{y^{(j)}} & j = 1, \dots, J \\
& z_X, z_Y \geq 0,
\end{aligned} \tag{1}$$

where $\|\cdot\|_1$ denote the 1-norm, vector e has all entries equal to one, $z_X := [z_{x^{(1)}}; \ldots; z_{x^{(I)}}] \in \mathbb{R}^I_+$ and $z_Y := [z_{y^{(1)}}; \ldots; z_{y^{(J)}}] \in \mathbb{R}^J_+$ are the non-negative vectors of errors of group X and Y. Observation $x^{(i)} \in \mathbb{R}^n$ is correctly classified if $0 \leq z_{x^{(i)}} \leq 1$, misclassified otherwise. Further, $\nu \geq 0$ is a user-defined penalty parameter allowing a trade-off between the margin maximization ($\|a\|_1$) and tolerating misclassification ($e^\top z_X + e^\top z_Y$). Once the starting hyperplane (a, γ) is obtained, it is shifted in order to determine hyperplane $H_1 := (a, \gamma - 1 + \omega_1)$ such that all points corresponding to patients in X lie on one of its side, and hyperplane $H_2 := (a, \gamma + 1 - \omega_2)$ such that all points corresponding to patients in Y lie on the opposite side. Finally, through line search hyperplane H_3 is identified lying between H_1 and H_2 and minimizing the overall number of misclassified patients.

2.2 Robust Formulation

To handle uncertainty in data features, we now consider a robust counterpart of model (1) with uncertainty sets in the form of hyperrectangles and hyperellipsoids (see [5]). We assume the uncertainty of every patient data $x^{(i)} \in X \subseteq \mathbb{R}^n$, $i = 1, \ldots, I$ to be represented by the uncertainty set $\mathcal{U}(x^{(i)})$. Equivalently for every $y^{(j)}$, $j = 1, \ldots, J$. Then, the robust counterpart of model (1) corresponds to the following optimization model:

$$\begin{aligned} \min_{a,\gamma,z_X,z_Y} \quad & \|a\|_1 + \nu(e^\top z_X + e^\top z_Y) \\ \text{s.t.} \quad & \max_{x \in \mathcal{U}(x^{(i)})} \left[a^\top x\right] \leq \gamma - 1 + z_{x^{(i)}} \qquad i = 1, \ldots, I \\ & \min_{y \in \mathcal{U}(y^{(j)})} \left[a^\top y\right] \geq \gamma + 1 - z_{y^{(j)}} \qquad j = 1, \ldots, J \\ & z_X, z_Y \geq 0. \end{aligned} \tag{2}$$

Uncertainty sets are built as follows:

- Let $\zeta_{x^{(i)}} \in \mathbb{R}^n_+$ define the perturbation vector of $x^{(i)}$ and let $\rho_X \in \mathbb{R}_+$ be the global measure of uncertainty for group X. Then, the hyperrectangular uncertainty set $\mathcal{U}_\mathcal{B}(x^{(i)})$ centered around $x^{(i)}$ is defined as:

$$\mathcal{U}_\mathcal{B}(x^{(i)}) := \left\{x \in \mathbb{R}^n \mid x^{(i)} - \rho_X \zeta_{x^{(i)}} \leq x \leq x^{(i)} + \rho_X \zeta_{x^{(i)}}\right\} \tag{3}$$

  and equivalently for every $y^{(j)}$, $j = 1, \ldots, J$.
- Let $\Sigma_{x^{(i)}} \in \mathbb{R}^{n \times n}$ be the positive definite covariance matrix associated to $x^{(i)}$. Then, the ellipsoidal uncertainty set $\mathcal{U}_\mathcal{E}(x^{(i)})$ centered around $x^{(i)}$ with ray $\rho_X \in \mathbb{R}_+$ is defined as:

$$\mathcal{U}_\mathcal{E}(x^{(i)}) := \left\{x \in \mathbb{R}^n \mid (x - x^{(i)})^\top \Sigma_{x^{(i)}}^{-1} (x - x^{(i)}) \leq \rho_X^2\right\} \tag{4}$$

  and equivalently for every $y^{(j)}$, $j = 1, \ldots, J$.

Once the starting hyperplane (a, γ) determined in (2) is obtained, it is shifted in order to determine hyperplane H_1, H_2 and H_3 as described above.

2.3 *Distributionally Robust Formulation*

We now consider a less conservative way to handle uncertainty in data features by considering a distributionally robust counterpart of model (1), where we treat all input patients data $x^{(i)}$, $i = 1, \dots, I$ as random variables for which the exact probability distributions $\mathbb{P}^{\text{true}}_{x^{(i)}}$, $i = 1, \dots, I$ are unknown (see [5]). For each $x^{(i)}$ we optimize against the worst-case expectation under all possible distributions $\mathbb{P}$ belonging to the ambiguity set $\mathcal{D}(x^{(i)})$. Equivalently for $y^{(j)}$ and $\mathcal{D}(y^{(j)})$. Accordingly, the distributionally robust counterpart of model (1) can be formulated as follows:

$$
\begin{aligned}
\min_{a,\gamma,z_X,z_Y} \quad & \|a\|_1 + \nu(e^\top z_X + e^\top z_Y) \\
\text{s.t.} \quad & \sup_{\mathbb{P}\in\mathcal{D}(x^{(i)})} \mathbb{E}_{\mathbb{P}}\big[a^\top x\big] \le \gamma - 1 + z_{x^{(i)}} \qquad i = 1, \dots, I \\
& \inf_{\mathbb{P}\in\mathcal{D}(y^{(j)})} \mathbb{E}_{\mathbb{P}}\big[a^\top y\big] \ge \gamma + 1 - z_{y^{(j)}} \qquad j = 1, \dots, J \\
& z_X, z_Y \ge 0.
\end{aligned}
\tag{5}
$$

We consider the general class moment-based ambiguity set proposed in [9] where the support and a list of partial moments describing the uncertainty are available:

$$
\mathcal{D}(x^{(i)}) := \left\{ \mathbb{P} \in \mathcal{P}^n_+ \;\middle|\; \begin{array}{c} \mathbb{P}\big(x \in \mathcal{U}_{\mathcal{B}}(x^{(i)})\big) = 1 \\ \mathbb{E}_{\mathbb{P}}\big[g_p(x)\big] \le (\varrho_X)_p \quad p = 1, \dots, n \end{array} \right\}
$$

with $\mathcal{P}^n_+$ representing the set of probabilities distributions on $\mathbb{R}^n$. Specifically, the first constraint requires every realization to be constrained within its support set $\mathcal{U}_{\mathcal{B}}(x^{(i)})$ defined as in (3), while the second group of constraints characterizes the moments information via n functions $g_p(\cdot)$, and enforces the generalized moment $\mathbb{E}_{\mathbb{P}}\big[g_p(x)\big]$ not to exceed a given threshold $(\varrho_X)_p \in \mathbb{R}_+$, $p = 1, \dots, n$. The moment function we employ in this paper is the piecewise linear formulation which can be interpreted as the first-order deviations of the uncertain parameter with respect to the nominal patient value $x^{(i)}$ along projections $f_X^{(p)} \in \mathbb{R}^n$: $g_p(x) := |f_X^{(p)\top}(x - x^{(i)})|$, $p = 1, \dots, n$. To determine projections $f_X^{(1)}, \dots, f_X^{(n)}$ and thresholds $(\varrho_X)_1, \dots, (\varrho_X)_n$ we adopt the database strategy based on *Principal Component Analysis* (PCA), described in [5]. Equivalently for every $y^{(j)}$, $j = 1, \dots, J$. Model (5) is intractable due to the infinite number of probability distributions contained in every ambiguity set; therefore, in [5], a tractable deterministic reformulation is provided.

Among the proposed formulations, model (2) with hyperrectangles uncertainty sets leads to the most conservative solutions. On the other hand, hyperellipsoids uncertainty sets lead to less conservative solutions, as those situations under which all features jointly assume extreme interval values are disregarded. Finally, the distributionally robust model (5) represents the most aggressive approach among the three, since it does not only rely on support information, but rather also assumes to have knowledge about the moments distributions.

3 Experimental Study

3.1 Data Collection

From an initial group of 150 COVID-19 patients admitted to the *Emergency Room* (ER) of Valcamonica Hospital (Esine, Brescia—Lombardy, Italy) between March 5th and April 1st 2020 and diagnosed with COVID-19, two different cohorts were selected for having complete clinical and laboratory data. The first cohort included 78 COVID-19 patients (47 males, 31 females, age range 32–90) who survived and could hence be discharged, whilst the second group encompassed 72 patients (20 males, 52 females, age range 55–95) who died within their hospital stay. All patients were diagnosed with COVID-19 according to current standards, i.e., displaying suggestive findings at chest *Computed Tomography* (CT; the classic ground glass pattern of interstitial pneumonia for a minimum of 35–40% of lung parenchyma) and positive results of real-time *Reverse Transcriptase Polymerase Chain Reaction* (RT-PCR) for SARS-CoV-2. Real-time RT-PCR was used for direct virus identification in nasopharyngeal swabs and was performed in reference laboratories from Lombardy network of COVID-19 diagnostics. For every patient, clinical history, signs, symptoms and results of laboratory investigations were collected from the local *Laboratory Information System* (LIS), as well as from medical records, using a standard form adopted for reporting data on infectious diseases to the Italian Ministry of Health. All clinical and laboratory data here described have been recorded upon hospital admission, and the quality of collection results was validated with *Internal Quality Control* (IQC) procedures and participation to the *External Quality Assessment Scheme* (EQAS) of Lombardy region, Italy. Clinical and laboratory information was gathered during clinical workout and the study was preliminary approved by the Ethical Committee of Brescia (certificate no. NP 4036). The study was carried out in accordance with the revised declaration of Helsinki and with the term of local legislation.

3.2 Numerical Investigation

Models of Sect. 2 have been trained using data of patients described above. First we perform a selection of input features, then we validate the models. The computations have been performed on a 64-bit machine with 8 GB of RAM, a 1.8 GHz Intel i7 processor, and numerical results are obtained under Matlab R2022a environment using MOSEK solver (version 8.1.0.72).

3.2.1 Feature Selection

The attributes considered are comorbidities, vital and laboratory parameters. Specifically, the following 26 predictors have been initially examined: test COVID-19 (positive, negative, doubted), sex (male, female), age (range 32–95), presence of chronic diseases (yes, no), presence of neoplasia (yes, no), diabetes (yes, no), presence of cardiovascular diseases (yes, no), immunodeficiency (yes, no), presence of respiratory diseases (yes, no), presence of kidney diseases (yes, no), presence of metabolic diseases (yes, no), body mass index BMI in the range of 30–40 (yes, no), body mass index BMI greater than 40 (yes, no), white blood cell count WBC (range 2, 18–25, 53), platelet count PLT (range 64–521), neutrophils (range 1, 31–21, 43), lymphocytes (range 0, 27–2, 77), D-dimero (range 270–20.000), aspartate aminotransferase AST (range 9–464), lactate dehydrogenase LDH (range 117–1.161), creatine kinase CK (range 7–4.038), C-reactive protein PCR (range 8, 3–372, 6), high sensitivity cardiac troponin I cTnI (range 5–5.124), ferritin (range 127–8.413), WBC/lymphocytes % (range 1, 78–46, 95) and emogas P/F (range 36–687, 14).

Since too many input attributes can cause overfitting and consequent poor performance of the ML algorithms, a χ-square test has first been performed, which allows reducing the number of input features by identifying whether each predictor is independent of the response variable (survived/deceased). According to the χ-square test, the following 15 attributes are the most important: cTnI, LDH, P/F, WBC/lymphocytes %, age, D-dimero, presence of chronic diseases, AST, neutrophils, diabetes, test COVID-19, PCR, presence of kidney diseases, CK and presence of respiratory diseases.

3.2.2 Models Validation

Models have been trained using a 50-folds cross-validation scheme: the data sample has been split into 50 sub-groups $i = 1, \ldots, 50$ each containing exactly 3 observations; the first subset is then used to validate the model that has been trained using the remaining groups $i = 2, \ldots, 50$. The process is repeated 50 times so that each subset i is used exactly once for validation. The goodness of the models is evaluated by measuring the indicators associated with group i:

$$\mathrm{A}_i := \frac{\mathrm{TP}_i + \mathrm{TN}_i}{\mathrm{TP}_i + \mathrm{TN}_i + \mathrm{FP}_i + \mathrm{FN}_i}, \quad \mathrm{R}_i := \frac{\mathrm{TP}_i}{\mathrm{TP}_i + \mathrm{FN}_i}, \quad \mathrm{P}_i := \frac{\mathrm{TP}_i}{\mathrm{TP}_i + \mathrm{FP}_i},$$

where TP_i stands for true positive, TN_i for true negative, FP_i for false positive and FN_i for false negative of group i. Finally, scores have been averaged as:

Table 1 Accuracy, recall, precision, and average CPU times of the deterministic, robust and distributionally robust SVM models. For each formulation, the best variant in terms of accuracy is highlighted in bold

	Deterministic (1)	Hyperrectangular Robust, (2)–(3)			Hyperellipsoidal Robust, (2)–(4)			Distributionally Robust, (5)		
$\rho_X = \rho_Y$	–	0, 1	0, 2	0, 3	0, 1	0, 2	0, 3	0, 1	0, 2	0, 3
Accuracy	77,33%	**80,00%**	79,33%	74,67%	78,66%	**78,66%**	77,33%	**77,33%**	75,33%	70,67%
Recall	63,89%	**63,89%**	63,89%	58,33%	63,89%	**65,28%**	63,89%	**59,72%**	59,72%	65,28%
Precision	85,19%	**92,00%**	90,20%	84,00%	88,46%	**87,04%**	85,19%	**89,58%**	84,31%	71,21%
Avg CPU times	2.318 sec.s	2.730 sec.s			2.768 sec.s			4.777 sec.s		

$$\text{Accuracy} := \frac{\sum_{i=1}^{50} \mathrm{A}_i}{50}, \quad \text{Recall} := \frac{\sum_{i=1}^{50} \mathrm{R}_i}{50}, \quad \text{Precision} := \frac{\sum_{i=1}^{50} \mathrm{P}_i}{50}.$$

For each formulation of Sect. 2, we report in Table 1 the indicator scores for increasing levels of robustness ρ_X, ρ_Y. For all formulations the user-defined penalty parameter $\nu = 1{,}5\mathrm{e}^{-2}$. For robust models, perturbation vectors $\zeta_{x^{(i)}}$, $\zeta_{y^{(j)}}$ and covariance matrices $\Sigma_{x^{(i)}}$, $\Sigma_{y^{(j)}}$ are set to be proportional to group X and Y standard deviations, while for the distributionally robust formulation moment thresholds are proportional to a scale factor K tuned to 0, 5 (see [5]).

Results show that the best performing model is the robust formulation with uncertainty sets having the form of hyperrectangles (2)–(3) and $\rho_X = \rho_Y = 0{,}1$ that records an overall accuracy of 80,00%; the low value of recall (63,89%) is related to a high value of false negative rate (36,11%), while the high precision score is due to a low false positive rate (5,13%). See Table 4. It is worth noticing that also the ellipsoidal robust model (2)–(4) can outperform its deterministic counterpart (1) for every choice of ρ_X, ρ_Y. On the other hand the distributionally robust variant (5) shows to be the weakest one: the model records the same accuracy level of the deterministic formulation, but while its precision increases, its recall reduces. This is coherent with what concluded in [5], according to which robust classifiers are especially beneficial for low dimensionality training sets (i.e., with less than approximately 500 observations), while as the training set gradually increases, features behavior is learned and higher levels of out-of-sample accuracy can be achieved via less conservative models (i.e., DRO formulations). Overall, with respect to deterministic approaches, [5] proves that more conservative methods, allow finding a trade-off between increasing the average performance accuracy and protecting against uncertainty.

Considering the 15 features mentioned above, other separation models from the literature were trained using the Matlab R2022a Classification Learner App, which allows to perform automated training to seek the best classification model among more than 20 possible choices, including: *Decision Trees* (DT), linear and quadratic *Discriminant Analysis* (DA), *Logistic Regression* (LR), *Naive Bayes* (NB), *Support Vector Machines* (SVM), *k-Nearest Neighbors* (k-NN) classifiers and *Ensemble Classification* (EC). Results are reported in Table 2, where we just show accuracy levels

Table 2 Accuracy levels of models from the literature trained using Matlab classification learner app

Accuracy (%)		Accuracy (%)		Accuracy (%)		Accuracy (%)	
Fine DT	77,30	Kernel NB	**80,00**	Fine *k*-NN	70,70	EC Boosted DT	76,70
Medium DT	77,30	Linear SVM	76,00	Medium *k*-NN	72,70	EC Subspace DA	78,70
Coarse DT	79,30	Quadratic SVM	76,00	Coarse *k*-NN	56,70	EC Subspace *k*-NN	65,30
Linear DA	76,00	Cubic SVM	76,00	Cosine *k*-NN	76,70	EC RUS Boosted DT	75,30
Quadratic DA	Failed	Fine Gauss. SVM	73,30	Cubic *k*-NN	72,00		
LR	76,70	Medium Gaussian SVM	**80,00**	Weighted *k*-NN	74,70		
Gaussian NB	Failed	Coarse Gaussian SVM	78,70	EC Bagged DT	75,40		

for the sake of readability. Robust formulation with hyperrectangles (2)–(3) is able to outperform all the reported models, achieving the same accuracy level of kernel NB, and medium Gaussian SVM (both of them non-linear models).

For every model, Matlab allows the internal hyperparameters tuning (e.g., the maximum number of splits for a DT). Additionally, instead of manually selecting these options, the Classification Learner App also provides the hyperparameter optimization to automate the selection of these values. For a given model type, the app tries different combinations of hyperparameter values by using an optimization scheme that seeks to minimize the model classification error, and returns a model with the optimized hyperparameters. See Table 3. Overall, results show that the strongest model in term of prediction accuracy is the *Optimizable DT* (ODT), that is, a simple classifier consisting of a sequence of binary decisions hierarchically organized. The ODT has an accuracy rate of 86,00%, a recall rate of 80,56% and a precision rate of 89,23%. Its prediction rate is around 1200 observations per second and its training time is 80,125 seconds. The maximum number of subdivisions considered is 6, with a subdivision criterion given by the Gini diversity index. The type of optimization considered is the Bayesian one.

As opposed to black-box modeling strategies which are typically difficult to illustrate, the ODT model benefits from the great potential of interpretability and the structure we obtained is reported in Fig. 1: to predict the answer (survival 0, mortal-

Table 3 Accuracies of optimizable models trained via the classification learner app

Accuracy (%)		Accuracy (%)		Accuracy (%)		Accuracy (%)	
Optimizable DA	77,30	Optimizable NB	80,00	Optimizable k-NN	81,30	Optimizable EC	83,30
Optimizable SVM	78,00	Optimizable DT	**86,00**				

Table 4 Confusion matrix for the hyperrectangular robust SVM formulation with $\rho_X = \rho_Y = 0,1$ (*up*) and confusion matrix for the ODT (*down*)

		Predicted class	
		0	1
True class	0	94,87%	5,13%
	1	36,11%	63,89%
		Predicted class	
		0	1
True class	0	91,00%	9,00%
	1	19,44%	80,56%

ity 1), we follow the decisions in the tree from the root node, down to a leaf node which contains the response. The results show that among the 15 attributes considered, the most relevant are cardiac troponin (cTnI), lactate dehydrogenase (LDH), aspartate aminotransferase (AST) and emogas (P/F). These results confirm doctors' observations. The root node considers the values of the cTnI attribute: if cTnI ≥ 21 and LDH $\geq 485{,}50$ the patient is unlikely to survive, otherwise he has good chances of recovery. The second most important attribute to consider is LDH, while AST and P/F appear at the final level of the decision tree. Note that, for the purpose of

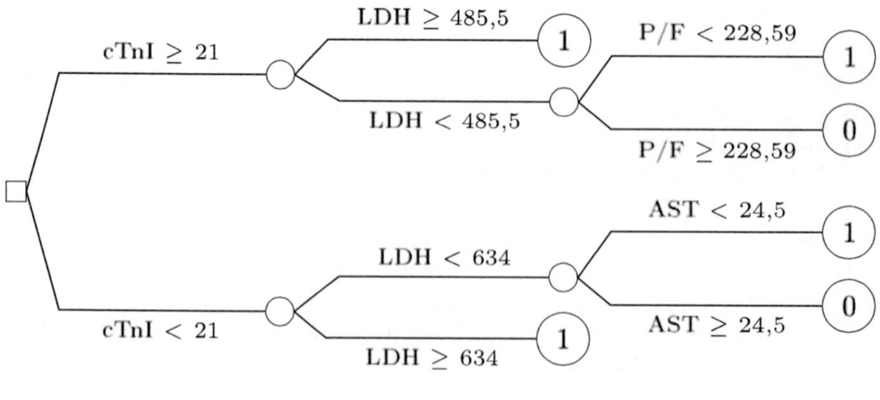

Fig. 1 The Optimizable Decision Tree

validation, a sensitivity analysis was also carried out with respect to an increasing number of input characteristics and to different cardinalities of the cross-validation subsets (5, 10, 15, and 30) confirming the same tree structure shown in Fig. 1.

4 Conclusions

This article presents novel data-driven optimization models to support doctors' decisions to solve one of the main problems encountered during the first months of the health emergency: predicting and calculating the risk of mortality from COVID-19, aiming at timely identify the most appropriate assistance, diagnostic and therapeutic cares for patients. To handle the uncertainty in data features, we formulated robust and distributionally robust classificators and compared their performance with other 25 different models from the literature. Results show that the best robust classifier is the one with hyperrectangles, which outperformed all the other 25 models and recorded an 80,00% accuracy level. However, considering optimizable models available using Matlab Classification Learner App, the best results were obtained with an optimizable decision tree, recording an 86,00% accuracy and allowing the identification of the most important predictors. Future works will then consider extending our robust and distributionally robust models for such a class of optimizable decision trees. Even if this tool has not been put into practice clinically, the results confirm doctors' experience. Therefore it could be used in the future for initial patients' assessments to accurately establish the severity of his/her condition, and enabling professionals to optimize accommodation by quickly identifying patients in need of intensive care.

References

1. Aloisio, E., Chibireva, M., Serafini, L., Pasqualetti, S., Falvella, F.S., Dolci, A., Panteghini, M.: A comprehensive appraisal of laboratory biochemistry tests as major predictors of COVID-19 severity. Arch. Pathol. Lab. Med. **144**(12), 1457–1464 (2020)
2. Ben-Tal, A., El Ghaoui, L., Nemirovski, A.: Robust Optimization, vol. 28. Princeton University Press (2009)
3. Bonetti, G., Manelli, F., Patroni, A., Bettinardi, A., Borrelli, G., Fiordalisi, G., Marino, A., Menolfi, A., Saggini, S., Volpi, R., et al.: Laboratory predictors of death from coronavirus disease 2019 (COVID-19) in the area of Valcamonica. Italy Clin. Chem. Lab. Med. **58**(7), 1100–1105 (2020)
4. CSG of the International Committee on Taxonomy of Viruses: The species severe acute respiratory syndrome-related coronavirus: Classifying 2019-nCoV and naming it SARS-CoV-2. Nat. Microbiol. **5**(4), 536 (2020)
5. Faccini, D., Maggioni, F., Potra, A.F.: Robust and distributionally robust optimization models for linear support vector machine. Comput. Oper. Res. **147**, 105930 (2022)
6. Huang, C., Wang, Y., Li, X., Ren, L., Zhao, J., Hu, Y., Zhang, L., Fan, G., Xu, J., Gu, X., et al.: Clinical features of patients infected with 2019 novel coronavirus in Wuhan. China Lancet **395**(10223), 497–506 (2020)

7. Lippi, G., Plebani, M.: The critical role of laboratory medicine during coronavirus disease 2019 (COVID-19) and other viral outbreaks. Clin. Chem. Lab. Med. **58**(7), 1063–1069 (2020)
8. Liu, X., Potra, F.A.: Pattern separation and prediction via linear and semidefinite programming. Stud. Inform. Control. **18**(1), 71–82 (2009)
9. Wiesemann, W., Kuhn, D., Sim, M.: Distributionally robust convex optimization. Oper. Res. **62**(6), 1358–1376 (2014)
10. World Health Organization: World Health Organization coronavirus disease (COVID-19) dashboard. World Health Organization (2020)

Integrating Decision Support Tools in the COD-19 Platform

Michele Barbato, Cristiano Carlevaro, Alberto Ceselli, Giuseppe Confessore, Gloria De Luca, and Marco Premoli

Abstract In early 2020 Northern Italy had to reorganize its whole public health system, in an extremely short amount of time to prevent collapse. A home health care service for mild COVID-19 patients (COD-19) was activated to relieve the pressure on hospital wards, while simultaneously improving the quality of service to these patients. The COD-19 service is designed to rely on a dedicated software platform, which serves for both operational support during emergencies and for strategic and tactical planning. We report our actions to provide institutional decision makers with advanced planning and operational tools which integrate the COD-19 platform. Three mathematical models allow for quantitative data analysis. They are designed to provide support both in a proactive way during early pandemic stages, and in a reactive way during and after emergency peaks. We also present experiments on real data from the COVID-19 pandemic emergency in Northern Italy, which indicate our approach to be effective.

Keywords Planning · Data analysis · Emergency management · COVID-19

M. Barbato · A. Ceselli (✉) · M. Premoli
Department of Computer Science, Università Degli Studi di Milano, Milano Via Celoria 18, 20133, Italy
e-mail: alberto.ceselli@unimi.it

C. Carlevaro
Spindox Spa, Via Bisceglie 76, 20152 Milan, Italy

G. Confessore
Consiglio Nazionale delle Ricerche, Area Territoriale di Ricerca di Bologna, Via Gobetti 101, 40129 Bologna, Italy

G. De Luca
ACT Operations Research IT Srl, Via Cavour 2, 22074 Lomazzo (CO), Italy

R. Aringhieri et al. (eds.), *Operations Research for Health Care in Red Zone*, AIRO Springer Series 10, https://doi.org/10.1007/978-3-031-38537-7_5

1 Introduction

Since the early stages of coronavirus pandemic, Lombardy Region in agreement with the Italian Ministry of Health activated response measures to manage the emergency and to limit the virus diffusion. The need for treatment of COVID-19 patients, that required specialized material and human resources, rapidly saturated public structures.

Our work stems in the context of the project Centro Operativo Dimessi COVID-19 (COVID-19 Discharged Operations Center) whose acronym is COD-19, supported by the Lombardy Region. It was developed and run as a response action in that emergency context and is currently in use. It concerns the design and adoption of a dedicated home health care service for mild COVID-19 patients. The overall aim of COD-19 is to relieve the pressure on hospital wards, improving at the same time the quality of service to mild patients.

To support health personnel in this ambitious objective, COD-19 relies on a dedicated information system, whose duty is to monitor and manage each home care patient. It consists of a virtual platform where all the data coming from a semi-automatic monitoring system is recorded and processed. The platform is meant to have both an operational and a tactical role. First, to support health personnel in its day-by-day duties, providing dedicated reports and alerts. Second, to act as a decision support system for the regional government, embedding planning tools for optimized health resource management.

The paper has two main targets: first, we report our actions to provide the institutional decision maker with advanced planning and operational tools to integrate the COD-19 platform. They are designed to operate at a logistics level, both in a proactive way during early pandemic stages, and in a reactive way to support emergency conditions during and after peaks. We present our findings from a methodological point of view, to foster the potential adoption of similar models in case preventing or at least limiting the spread of diseases is critical for population. As second target, being our approach quantitative, we report on our experiments on real data from the COVID-19 pandemic emergency in Northern Italy.

2 Proposed Approach

In the context of COD-19 virtual platform, we envisage a decision support system consisting of three main components.

Clustering and Policy Assignment (CPA): a tool for managing lockdown restrictions. It generates partitions of a target region into zones, allowing for different actions to be performed in different zones. Its aim is to help limiting or delaying the spread of the disease, introducing at the same time as few restrictions as possible for the population. It considers as input (1) topological data of a certain geographical area composed by several locations, some of which are adjacent and therefore share a border (2) epidemiological data. Its core are mathematical models that minimize a given measure of the epidemic spread by determining simultaneously the two following types of decisions under budget constraints: (a) to close some of the borders between adjacent locations (b) to assign intervention policies and restrictions to the sub-areas emerging from the border closing decisions.

Health Resources Management (HRM): dealing with massive logistics mobility. The treatment of COVID-19 patients requires specialized material and human resources, often exceeding the available amount and hence requiring re-allocation of health resources in a limited time. The HRM embeds mathematical models to support such re-allocation among hospitals in a territory with two goals: first, to minimize worsening of quality of service for patients, and second to keep re-organization costs as low as possible. The HRM takes as input (1) the amount of resources and inpatients in each hospital of the region (2) a forecast on the peak of COVID-19 patients that will require treatment. The re-allocation yielded by HRM contains: (a) the location of COVID-19 wards to open, (b) the plan to transfer resources and patients among wards and hospitals, (c) the number of patients that will be discharged or for which the hospitalization will be postponed and, finally (d) an estimate of the additional resources that need to be found to achieve a target quality of service (either purchased, manufactured or borrowed from other territories).

Disease Course Predictive (DCP): tool for predicting disease outcomes of each patient. The DCP takes the form of a software platform which elaborates data from the monitoring system in real-time, trains predictive models of the disease course from historical data and makes them accessible through an interactive form. Specifically, the health personnel can insert personal information, medical and personal data of a patient (e.g., age, sex, chronic diseases) obtaining a prediction on their disease outcome, thereby helping to decide whether to hospitalize, treat with home health care or simply discharge the patient.

Decision support system: Figure 1 reports a schematic view of our approach. A use-case in the context of health emergencies can be the following. When the epidemic situation is at the beginning, the decision maker starts a strategic planning of possible containment policies. They get initial estimates on the epidemiological dynamics of the disease, as well as retrieving topological data of the territory. Then, they feed the CPA with policies, epidemiological and topological data. The CPA produces a quantification on the effect of containment policies when clustering is done in an optimal way, together with an actual clustering solution. The decision maker then revises the containment policies and refine the epidemiological data, and possibly iterate the process. At this stage, they can use the HRM for scenario analysis in selected clusters. This provides (1) a check on the feasibility of the scenario in terms of minimum level of quality of service and (2) early insights on which prior arrangements between hospitals would be needed if the scenario realizes.

Later, if an epidemic outbreak arises, the decision maker may decide to actually implement the containment measures designed in the previous strategic phase. In this second phase they use the HRM as a tactical planning tool, actually producing plans for relocating resources, as well as guidelines for rerouting people in case of congestion.

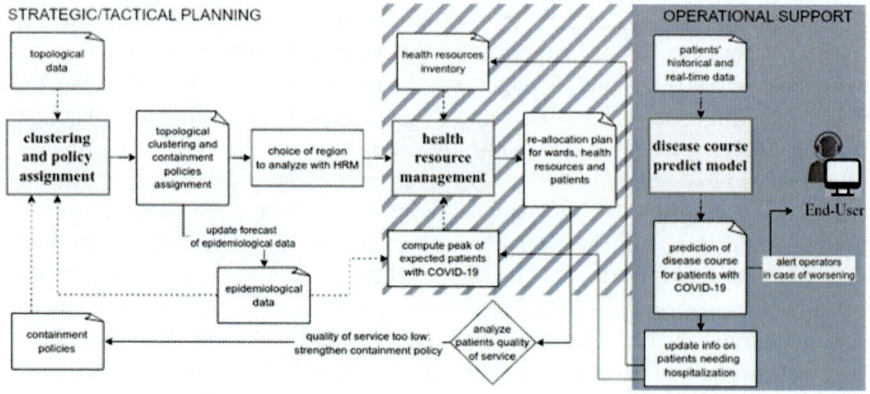

Fig. 1 Schematic view of the proposed approach

Finally, during the emergency peak, the decision maker uses information systems embedding the DCP tool for daily operations. Information collected by the DCP can also be used to improve data accuracy, and re-run the HRM to possibly revise relocation and re-routing strategies on the fly (as long as the emergency leaves the flexibility to do so).

After this stage, only the DCP keeps helping health personnel in day-by-day cares to patients, for instance during triage and home health care management.

In the following we detail the inner structure of each tool, in terms of formal methods and system design techniques employed.

2.1 Clustering and Policy Assignment

The CPA tool models the target geographical area as a graph G whose vertices represent locations and whose edges represent borders between adjacent locations—see Fig. 2 (left). The flexibility of this graph-theoretical model lies in the use of vertices representing locations of distinct administrative division levels, e.g., provinces, single municipalities, groups of municipalities, etc.

Each vertex of the graph carries information about its epidemic state at an initial instant $t = 0$, identified with its number of susceptible $S(0)$, infectious $I(0)$ and recovered $R(0)$ individuals. The CPA tool aims to minimize a given measure of the epidemic spread by simultaneously (a) closing borders between adjacent locations in the geographical area, creating clusters and (b) assigning containment policies to the locations of each cluster. We assume the set K of policies to be pre-determined and already available, so that only their assignment to locations must be decided. Decisions (a) and (b) mitigate both disease propagation and adverse effects of containment policies on the population.

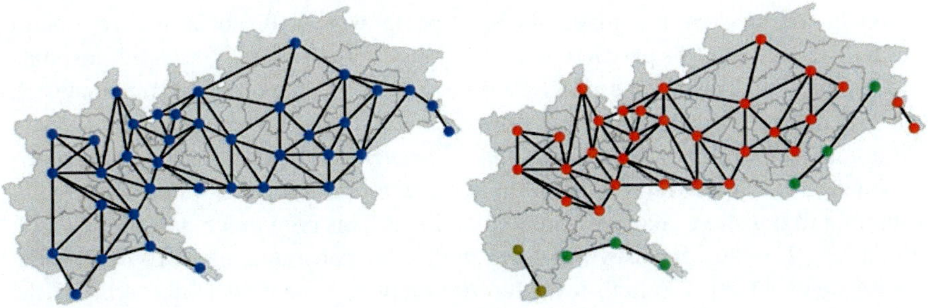

Fig. 2 Graph modelling of provinces used in the CPA tool (left) and clustering-assignment solution obtained via our optimization algorithm (right)

CPA variables, constraints and parameters. Variables $x_e \in \{0, 1\}$ correspond to decisions (a) by modelling removal of edge e from the graph. Variables $z_{ik} \in \{0, 1\}$ correspond to decisions (b) by modelling policy-vertex assignment. A high-level description of the optimization model underlying CPA is

$$\begin{aligned} \min\ & W(\beta, S(0), I(0), R(0), K) \cdot z(V \times K) \\ \text{s.t. } & x(E) \leq B_{\text{removal}}, \\ & C_{\text{policy}}(K) \cdot z(V \times K) \leq B_{\text{policy}}, \\ & x \in \textsc{ConnectedSubgraph}(F_k : k \in K), \\ & (x, z) \in \textsc{ConsistentAssignment} \end{aligned}$$

where $W(\cdot)$ is a time-independent function approximating the disease propagation via the compartmental Susceptible-Infectious-Recovered (SIR) model [6] ($W(\cdot)$ depends on the boundary conditions $S(0)$, $I(0)$, $R(0)$ and on parameter β representing the contact rate, which we assume to be known); B_{removal} is a given upper-bound on the number of removable edges; B_{policy} is a given upper-bound on the societal and economical cost due to containment policies, whose unitary costs are given by parameters C_{policy}; ConnectedSubgraph is the set of connected subgraphs of G with at least F_k vertices for clusters of type $k \in K$; ConsistentAssignment is the set of "coloured" subgraphs whose vertices are assigned the same colour. Solving the above model yields the best combination of decisions (a) and (b) which minimizes the total number of individuals transitioning from the susceptible compartment to the infectious one. The model is cast to a mixed-integer linear program (MILP), solvable by means of well-established algorithms.

2.2 Health Resources Management

The HRM model contains entities representing hospitals wards, health resources, and patients. Each model entity abstracts real-world entities in one-to-one, many-to-one or one-to many way. A ward entity is characterized by its medical specialty, possibly encoding a set of equivalent ones. Special ward and specialties are used to encode dischargement of patients, temporary support structures like field hospitals and external suppliers of resources. Resource entities include both staff and material. Each patient entity is identified by a required medical specialty and by the severity of illness. *HRM variables, constraints and parameters.* At its core, the mathematical model of HRM is a facility location allocation problem with multi-commodity resources and transfer of capacity among facilities. This core model is enriched with multiple details to match real-world needs. Full details on mathematical MILP model of HRM are given in [1]. A high-level description of the core model underlying HRM is the following:

$$\begin{aligned} \min\ & C^{QoS}(\pi) + C^{E}(\chi, \rho) \\ \text{s.t. } & \chi \in \textsc{SpecialtyAssignment}, \\ & (\pi, \chi) \in \textsc{PatientsAllocation}, \\ & (\rho, \chi) \in \textsc{ResourceTransfer} \\ & (\pi, \rho) \in \textsc{DemandCovering} \\ & t(\chi, \pi, \rho) \le T \end{aligned}$$

Variables $\chi_{w,s} \in \{0, 1\}$ model the assignment of specialty $s \in S$ to ward $w \in W$, that must comply with policies SPECIALTYASSIGNMENT. Variables $\pi_{p,w1,w2} \in \mathbb{Z}_{\geq 0}$ model the allocation of patients of type $p \in P$ from their original ward $w1 \in W$ to a destination ward $w2 \in W$. This latter may represent the 'dischargement' of the patient. This allocation must comply with policies PATIENTSALLOCATION. These latter may include the condition of the patient and the type or distance of destination ward. Variables $\rho_{r,w1,w2} \in \mathbb{Z}_{\geq 0}$ model the trasfer of resources of type $r \in R$ from their original ward $w1 \in W$ to a destination ward $w2 \in W$. The former may represent an external supplier. This transfer must comply with policies RESOURCETRANSFER. Patients require multiple types of resources to be available in the ward in which they are allocated, encoded in DEMANDCOVERING. Finally, $t(\cdot)$ measures the time required for each decision, which is limited by an upper bound T.

HRM key performance indicators. The evaluation of the re-allocation plan resulting from the HRM is composed by two elements. First, the quality of service C^{QoS} is evaluated using the number of inpatients that are moved among hospitals, and that are either discharged or delayed in treatment. Second, logistics effort C^{E} is measured by the costs and time of set-up of ward repurposing and resource transportation. The search for decisions optimizing KPIs, while respecting feasibility constraints, is formulated as a MILP and solved by an ad-hoc mathemtical programming heuristic presented in [1]: the heuristic exploits MILP decomposition methods and column generation algorithms, which in turn drive a very large scale neighborhood search. A bi-objective optimization algorithm is also presented to compare the two KPIs C^{E} and C^{QoS}

2.3 Disease Course Prediction

It has been developed a model that can be used to predict the COVID-19 disease progression from the initial patient history and symptoms. Its aim is to allow physicians to make better decisions for the patient as well as the use of health resources within their hospitals. Data describing patient's history and symptoms were analysed, alongside the patient's disease course and outcome using three different types of mathematical model. Linear models, artificial neural network (ANN) and convolutional neural network models have been evaluated. Preliminary results have showed that ANNs are better in adapting to the COD-19 type of data. Even if long training times are usually reported in the literature for the ANN, they are still suitable to the COD-19 setting. Prediction times are instead similar among linear models and ANNs, and therefore both are suitable for the real-time task of the virtual platform.

3 Case Study

To illustrate the functioning and effectiveness of our decision-support tool, we report results on data drawn from the 2020 COVID-19 emergency in Northern Italy. They aggregate diverse sources.

First, we had access to rich data about 1772 COVID-19 patients monitored in a time range spanning about six months (February 2020–August 2020), Their level of disease was vary: mild ones correspond to those treated by home health care services, more severe ones are hospital inpatients. These data contain sensitive information, such as medical evaluations, and is therefore available only to the partners of the COD-19 project. Second, open data sources have been employed representing the timeline of COVID-19 pandemic in Italy and the availability and usage of health resources in the 38 Northern Italian provinces [2, 4, 5]. Additional open data sources have been used to generate the usage of health resources in 86 hospitals in Lombardy region in Italy before COVID-19 emergency [7], which have been used to set up experiments for the HRM tool.

3.1 Clustering and Policy Assignment: Design and Results

To experiment with the CPA tool, we consider provinces as locations. The graph modelling the corresponding geographical area is given in Fig. 2. We consider three available sets of policies of increasing level of restriction. They are indicated by green, yellow, and red colours.

The model parameters concerning the total population of the locations and the number of corresponding infectious individuals are retrieved by the open-source repositories reported at the beginning of this Section. Letting $\tau = 14$ be the average

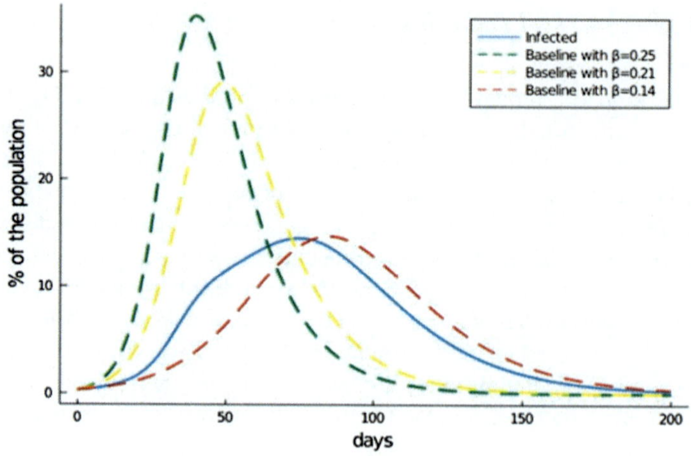

Fig. 3 Solution quality assessment via SIR simulation. Only the evolution of the infectious people is plotted for each scenario

number of days during which an individual stays infectious, we set $\gamma = 1/\tau$ (infectious rate) and $\beta = R0/\tau$ (avg. contacts per person per time), where $R0$ is the basic reproductive number as reported in the literature [3], adjusted according to the containment policy. We obtain $\beta = 0.25, 0.21, 0.14$ for the green, yellow and red clusters respectively. Following the notation of Sect. 2.1, we set $B_{\text{removal}} = 19$ ($\sim$21.6% of the total edge number) and $B_{\text{policy}} = 91\%$MPC where MPC is the maximum possible cost, obtained by multiplying the cost of the strictest policy by the total susceptible population. The solution obtained after optimization is subsequently evaluated with a SIR model adapted to our tool.

Optimization and simulation software are written in C++ using CPLEX 12.6.3 as MILP solver. The resolution of the optimization model leads to the solution depicted in Fig. 2 (right). It has five clusters, two of which are assigned the mildest policy and obtained by removing a total of 16 edges. The assessment of the solution quality is shown in Fig. 3.

The result is displayed along with the expected infectious evolution in some baseline scenarios, in which all borders are shut down (full lock-down) and all locations are assigned a same policy. For the solution in Fig. 2 (right) the corresponding SIR-based simulation plot yields the solid blue line, corresponding to the expected evolution of the overall infectious compartment, if the proposed solution is implemented in practice. The dashed lines represent infectious people in the baseline scenarios (e.g., red dashed line is full lockdown on all region). Our solution and the strictest-policy scenario yield both a peak of infectious individuals of about 15% of the total population after about three months since the epidemic start (we remind data is cumulative); the main difference is that ours reaches the peak slightly before and faster, but allows to save about 9% of containment restrictions.

We have also tested the CPA tool over a set of 110 instances obtained by combining 11 values of $B_{removal} \in \{11, 15, 19, \ldots, 47\}$ (that is, from 11 to 47 increasing by 4) and 10 values of $B_{policy} \in [0.65\text{MPC}, 0.91\text{MPC}]$ (values equally distributed) and in which $|K| = 3$, $F_k = 3\ \forall k \in K$, and $C_{\text{policy}}(2) = 1.3C_{\text{policy}}(1)$ and $C_{\text{policy}}(3) = 2C_{\text{policy}}(1)$. With a time limit of 1200 CPU seconds, we obtain optimal solutions for 46 instances. On the solved instances, the median resolution time is of 115.06 CPU seconds, with a maximum of 1191.44 and a minimum of 9.78 CPU seconds. The average resolution time is of 292.68 CPU seconds. The median relative optimality gap on the remaining 64 instances is of 12.46%, with a maximum of 29.34% and a minimum of 0.06%. In average the relative optimality gap is 13.10%. The difficulty of the instances is most influenced by the ratio $B_{\text{removal}}/|E|$ (larger values corresponding to more difficult instances).

3.2 Health Resources Management: Design and Results

Concerning the HRM module, we test on 86 hospitals of Lombardy, covering more than 10 million inhabitants. The model is populated by public repositories described in [7]. The model includes four many-to-one ward entities (mild and severe COVID and non-COVID) and four one-to-one entities (traumatology, neurosurgery/neurology, cardiology, non-COVID Intensive Care Unit). These match Lombardy policies on essential health services. All wards are considered as eligible to change specialty. Beds, physician, nurses and ventilators are considered as resources. Cost and time for the transportation of resources and patients are estimated to be directly proportional to the distance between hospitals. No limit is set on the time available to complete all transportation. The cost of a patient discharge is set to a very high value: no patient is discharged unless strictly needed for feasibility.

We compared the allocation plan resulting from our optimization with a naive initial re-organization in which COVID-19 wards are in the areas of the territory with highest incidence of the infection in the period February–April 2020 (red points Fig. 4a). The optimization of the allocation of resources starting from this naive solution results in the relocation of COVID-19 wards (red points in Fig. 4b). The optimized re-allocation provides savings of 45% of transportation costs with respect to the naive solutions, and allows to hospitalize 42% more patients, thereby highlighting how useful HRM would be in helping the health system to face future emergencies. As concerns computational requirements of HRM, Table 1 presents execution time, in seconds, and optimality gap for instances with increasing number of wards, averaging over 8 instances. Although HRM takes over 10 h for the biggest instances, such time is suited for a strategical/tactical plan for epidemics lasting weeks or months. Higher times results in lower optimality gap. Detailed computational analysis are given in [1].

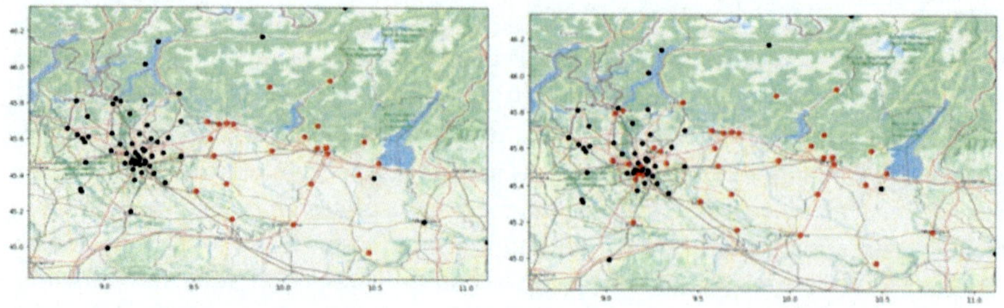

(a) solution driven by incidence of infection in the period Feb-Apr 2020 (b) solution optimized with Health Resources Management Tool

Fig. 4 Health resources management in lombardy. Dots represent hospital locations. Red dots represents hospitals in which at least one COVID-dedicated ward has been located

Table 1 Computational results of HRM

Number of wards	62	147	272	392	483
Time (s)	112.75	1276.88	4853.00	12553.00	38053.25
Optimality gap	16.76%	12.94%	5.71 %	3.82%	2%

3.3 Disease Course Prediction: Design, Integration and Results

The research output of the DCP is twofold. A web software platform was developed, providing a software dashboard for hospital personnel. Relevant information is presented to the final user by means of either numerical KPIs or charts. Using its descriptive statistics features we could rebuild, for instance, correlation measurement between hospitalization and age, existing chronic diseases, body mass index. We indeed found them, on our dataset of 1772 COVID-19 patients, to match well those computed on a national basis. On the modeling side, each patient was labeled in terms of length of stay at the hospital as follows: mild (0–1 days), average (2–3 days), severe (4–7 days), extreme (more than 7 days). Then, given patient features, we tried to predict their severity class. Being real data collected during the emergency, our dataset had several missing values. Nevertheless we tried, after data processing, to train supervised classification models using Python and the scikit-learn package. We compared linear predictors. The best one, based on logistic regression, could correctly classify 44% of the patients.

4 Conclusions

From a methodological point of view, we found machine learning, mathematical optimization models and software engineering to nicely fit one another in building effective decision support tools. Our case study reveals our approach to be viable and promising also in real world contexts. Indeed, by allowing to efficiently share relevant data, our tools may help both health operators in day-by-day operations during emergency peaks, and institutional decision makers in proactively evaluate tactical plans in early stages of outbreaks. Finally we remark that, although our work was conceived in the context of the COVID emergency, the models, methods and process protocols we devised are *not* strictly related to this specific disease. Rather, they can be applied in case of any (actual or potential) epidemic diffusion, to help facing the emergency at best.

Acknowledgements This work was partially supported by the Italian Lombardy Region, funded by Regional Operative Program—European Regional Development Fund 2014-2020 with Grant ID R1.2020.0002349, project COD-19.

References

1. Barbato, M., Ceselli, A., Premoli, M.: On the impact of resource relocation in facing health emergencies. Eur. J. Oper. Res. (2022). https://doi.org/10.1016/j.ejor.2022.11.024
2. Dipartimento della Protezione Civile, P.: COVID-19 Italia—Monitoraggio situazione (2020). https://github.com/pcm-dpc/COVID-19. Accessed 30 Jan 2021
3. Gatto, M., Bertuzzo, E., Mari, L., Miccoli, S., Carraro, L., Casagrandi, R., Rinaldo, A.: Spread and dynamics of the covid-19 epidemic in italy: Effects of emergency containment measures. Proc. Natl. Acad. Sci. **117**(19), 10484–10491 (2020)
4. Task force COVID-19 del Dip. Malattie Infettive e Servizio di Informatica, I.S.d.S.: Epidemia COVID-19, Agg. nazionale: 13-01-2021. https://www.epicentro.iss.it/coronavirus/bollettino/Bollettino-sorveglianza-integrata-COVID-19_13-gennaio-2021.pdf. Acessed 30 Jan 2021
5. Istat: Popolazione residente al 1 gennaio (2021). http://dati.istat.it/Index.aspx?DataSetCode=DCIS_POPRES1. Accessed 17 March 2021
6. Martcheva, M.: An Introduction to Mathematical Epidemiology, vol. 61. Springer, Berlin (2015)
7. Premoli, M., Barbato, M., Ceselli, A.: COVID-19 Hospital Resource Management Dataset—Lombardy 2020—Replication Data (2021). https://doi.org/10.13130/RD_UNIMI/WWUZIJ

A Semi-online Ambulance Routing and Scheduling Problem with Complex Patient-Vehicle Relations

Julia Resch

Abstract We consider a new complex variant of a dial-a-ride problem arising from the daily operations of the Austrian Red Cross. This non-profit organization faces the challenge that its vehicles do not only provide emergency services, but also patient transfers for non-emergency medical appointments. For the latter, some transportation requests are known well in advance, while others arrive at relatively short notice during the day. In addition to this semi-online scheduling setting, practical constraints and operational features are considered. The main aspects are multiple depots, a mixture of short and long time windows, a heterogeneous vehicle fleet allowing for various loading patterns, and patients with different priorities and transportation mode requirements. We aim to design routes and schedules in order to serve all transportation requests of a day with a given fleet of vehicles while maximizing the overall service level associated with the deviations of patients' actual arrival and/or departure times from their desired ones. For this problem, we present a constructive heuristic that follows a selective insertion procedure. Computational results on real-world data show that the proposed heuristic is well suited for coping with the difficulties of the problem in practice.

Keywords Dial-a-ride problem · Patient transportation · Ambulance management · Heterogeneous fleet · Heuristic

1 Introduction

The dial-a-ride problem (DARP), an extensively studied problem in the area of operations research, is a pickup and delivery problem in which a fleet of vehicles has to serve a given set of user-specified transportation requests, each defined by an origin and a destination (see, e.g., [3]). Over the last decades, a vast number of variants of the DARP has been introduced in the literature. For a comprehensive survey, we

J. Resch (✉)
Department of Operations and Information Systems, University of Graz, Universitaetsstrasse 15, 8010 Graz, Austria
e-mail: julia.resch@uni-graz.at

R. Aringhieri et al. (eds.), *Operations Research for Health Care in Red Zone*, AIRO Springer Series 10, https://doi.org/10.1007/978-3-031-38537-7_6

refer to [3], state-of-the-art reviews of recent developments and solution approaches are provided by [6, 7].

In this paper, we investigate a new variant of the DARP with multiple depots and heterogeneous vehicles and users which is motivated by a problem faced by the Austrian Red Cross in its daily operations. This non-profit organization provides ambulance services in a hybrid system by serving both emergency services and non-emergency door-to-door patient transfer services. For the latter, some transportation requests are known well in advance, while others arrive at relatively short notice during the day. Instead of studying the DARP in its online version, we are interested in a *semi-online* variant, in which we assume that all requests are known, but we have the restriction that vehicles cannot start towards the requested pickup before the request was actually announced. In fact, there are several reasons for employing this setting: (i) We determine ex-post the best vehicle composition for each day of the data set by running the heuristic on numerous vehicle fleet compositions. This allows an efficiency analysis of the general planning procedures. (ii) Based on this information and on the requests announced so far, one may predict the number of vehicles needed for the next day. (iii) The optimized deployment plans can be used to evaluate and improve the performance of the dispatcher by an analysis of the schedule at the end of the day. (iv) By testing various fleet compositions over a longer time horizon, one may gain information about the fleet deployment and purchases of new vehicles. It should be pointed out, however, that we are not interested in a decision tool for the online scheduling task of the dispatcher, although our solution approach could easily be adapted for this purpose.

From the modeling perspective, the considered problem takes into account the following new aspects in addition to the already mentioned semi-online feature:

(i) a heterogeneous fleet of vehicles and patients with different requirements for the mode of transportation, implying compatibility constraints between vehicles and patients and resulting in different loading patterns for the joint transport of patients;
(ii) priorities assigned to patients affecting the length of the time windows and limiting the set of vehicles that are capable to carry out a patient's transfer;
(iii) allowed waiting times for empty vehicles on the route (outside the depot) of at most 15 min.

While the existing literature on DARPs usually focuses on minimizing the provider's operating costs, e.g., total routing costs, our goal is the maximization of the overall service level associated with the deviations of actual arrival and/or departure times of patients from their requested times. For this purpose, a non-decreasing step function of deviations is introduced, taking into account that different priorities reflect different service levels.

Although DARPs can be solved exactly by various solution procedures such as branch-and-cut algorithms [2, 8] and branch-and-cut-and-price algorithms [1, 5], these exact approaches are limited to small-sized problem instances, and thus are ineligible for our large-scale instances. Beyond that, the considered problem incorporates a large number of constraints to cover numerous aspects from the real-world perspective, so even the construction of a feasible solution brings its own challenges. For this reason, we provide two variants of a constructive heuristic algorithm that are based on a simple but effective selective insertion procedure in which patients are assigned to vehicles. The variants differ, though, in how the requests are selected. In a computational study, both approaches are evaluated on real-world instances from an Austrian rural region.

The remainder of this paper is organized as follows. In Sect. 2 we present a detailed description of the problem under consideration. Sect. 3 is devoted to the main contribution of this work: the description of the proposed selective insertion heuristic. Preliminary computational results are reported in Sect. 4. Finally, Sect. 5 concludes the paper and outlines directions for future research.

2 Problem Description

In this section, we present our problem setting, which aims to find a set of routes for a given fleet of K heterogeneous vehicles in order to satisfy all R transportation requests while minimizing patient inconvenience.

Denoting by $\mathcal{I}$ the set of different vehicle types, the considered vehicle fleet $\mathcal{K}$, $|\mathcal{K}| = K$, comprises a partitioning $\mathcal{K} = \bigcup_{i \in \mathcal{I}} \mathcal{K}_i$, where $\mathcal{K}_i$ represents the subset of vehicles of type $i \in \mathcal{I}$.

Each vehicle $k \in \mathcal{K}$ has a total capacity Q_k which indicates the maximum number of patients to be transported simultaneously. Additionally, the vehicles are heterogeneous in terms of their capacity regarding different modes of transportation. Let Q_k^m denote the capacity of mode $m \in \{W, C, S\}$ on vehicle k where $m = W$ indicates a seated transport (for patients able to walk), $m = C$ a transport in a wheelchair or on a carrying chair, respectively, and $m = S$ a transport on a stretcher. In this context, it should be emphasized that the heterogeneous fleet of vehicles results in different loading patterns for the joint transportation of patients. Apart from this, there are multiple geographically distributed depots $\mathcal{H}$ at locations ℓ_h, $h \in \mathcal{H}$. Each vehicle k is associated with a depot $h_k \in \mathcal{H}$, from which it starts its route. In addition, each vehicle k has a fixed operating time window $[E_k, L_k]$, i.e., the vehicle is not allowed to leave its assigned depot earlier than E_k and the vehicle's last delivery must not take place later than L_k. Note that the vehicle has to return to its depot after the last delivery, but contrary to the classic DARP, this can take place outside the operating time window.

With the described vehicle fleet, a set of transportation requests $\mathcal{R}$, $|\mathcal{R}| = R$, has to be served. Each request $r \in \mathcal{R}$ incorporates a pickup r_p and a delivery r_d which take place at locations ℓ_{r_p} and ℓ_{r_d}, respectively. Let $\mathcal{L}$ indicate the set of all occurring locations, i.e., $\mathcal{L} = \{\ell_{r_p} \mid r \in \mathcal{R}\} \cup \{\ell_{r_d} \mid r \in \mathcal{R}\} \cup \{\ell_h \mid h \in \mathcal{H}\}$. For the sake of simplicity, we suppose that the shortest travel times t_{ij} between each pair of locations $i, j \in \mathcal{L}$ are known and constant. For each request $r \in \mathcal{R}$, pickup and delivery are restricted to occur within specified time windows where $w_{r_p} = [e_{r_p}, l_{r_p}]$ and $w_{r_d} = [e_{r_d}, l_{r_d}]$ represent the time windows for pickup and delivery, respectively.

The length of the time windows depend on the priority of a request, $\pi_r = \{low, medium\text{-}low, medium\text{-}high, high\}$, and on the shortest travel time between the depot closest to the requested pickup location and the pickup location itself. Although we have to serve all transportation requests which entails that they cannot be rejected, time windows are considered to be hard. Consequently, if any of the time window constraints cannot be met, there is no feasible solution to the problem with the given fleet of vehicles. In terms of priorities, a high priority is associated with an emergency, while priorities equal to low, medium-low or medium-high correspond to non-emergency transfers. Furthermore, for each request, there is an announcement time a_r which expresses the time that request r becomes known. In order to compute schedules which are comparable to online plans resulting from a human dispatcher, we impose the semi-online condition that a vehicle cannot leave its current position to drive towards pickup location ℓ_{r_p} earlier than a_r.

The considered requests concern patients who may require one of three different transportation modes, either a seated transport ($q_r = W$), a transport in a wheelchair or on a carrying chair ($q_r = C$), or a lying transport on a stretcher ($q_r = S$). The service time required to load (unload) a patient at the pickup (delivery) location is given by s_{r_p} (s_{r_d}) and depends on the patients' mode of transportation. Furthermore, patients may have special requirements that cannot be handled by the equipment in all types of vehicles. To address these limitations, we introduce patient-vehicle compatibility indicators $\delta_{rk}, r \in \mathcal{R}, k \in \mathcal{K}$, which take value 1 if request r can be transported by vehicle k, and 0 otherwise. Besides, requests may be scheduled to be served jointly by the same vehicle, meaning that multiple patients can be in a vehicle at the same time. However, the medical condition of a patient, reflected by the priority π_r, may not allow sharing the vehicle with other patients. These requirements can be modeled based on sharing indicators $\lambda_r, r \in \mathcal{R}$, which are 1 if request r permits sharing and does not require a single transport, and 0 otherwise.

The considered DARP consists of designing a set of vehicle routes such that:

- every route of a vehicle k starts from its assigned depot h_k;
- each pickup and delivery location is visited exactly once, each pickup and its associated delivery are served by the same vehicle and ℓ_{r_d} is visited later than ℓ_{r_p};
- the arrival at/departure from a location occurs within the specified time windows and the vehicle cannot leave for pickup before the request is announced;
- the utilization of the vehicles does not exceed their capacity at any time;
- the constraints on the compatibility between patients and vehicles and for joint transports are satisfied;
- a vehicle, as long as there is no patient on board, can wait up to 15 min at a delivery location before proceeding its tour (to avoid early arrival), or it returns to its assigned depot from where it can start a new route.

The objective is to maximize the overall service level associated with the deviations of actual arrival and/or departure times of patients from their desired ones. In fact, there are requests with an agreed appointment for e.g., medical examinations or therapies and those without. In the former case, it is essential that the appointment is met. Accordingly, in case of a late arrival, i.e., an arrival after the agreed time, the deviation should be kept to a minimum. For the latter case, however, the earlier the actual pickup takes place and the shorter the ride time, the better the service level for the patient. Besides, the situation is different for emergencies where it is crucial that a vehicle arrives at the scene of accident as soon as possible after receiving the emergency call. In order to express the service level in terms of user inconvenience, we introduce a non-decreasing step function, called service cost function, that maps the actual arrival and departure times, respectively, to a non-negative value. Let c_r indicate the service cost of request r consisting of the service costs for pickup and delivery, c_{r_p} and c_{r_d}. For notational convenience, we set the service cost for pickup to zero if it concerns a request with an appointment; similarly, the delivery service cost for emergency transports equals zero. Finally, the objective is to minimize the *total service cost*

$$c = \sum_{r \in \mathcal{R}} c_r = \sum_{r \in \mathcal{R}} \left(c_{r_p} + c_{r_d} \right).$$

3 A Selective Insertion Heuristic

For the stated problem, an integer linear programming (ILP) formulation based on a time expanded network was introduced in [4]. However, preliminary experiments have shown that due to their size our real-world instances cannot be solved to optimality in reasonable computing times, even not by warm starting the ILP with a feasible solution obtained from a heuristic. Therefore, this exact approach is not applicable in practice. Hence, in this section we focus on a simple but effective insertion heuristic in order to generate acceptable solutions of good quality within a reasonable amount of time. Basically, the proposed heuristic attempts to create routes and schedules by successively selecting a request and assigning it to the vehicle associated with the minimum increase in the objective value. A pseudocode description of our heuristic is given in Algorithm 1.

The heuristic takes as input an instance of the routing and scheduling problem and a given fleet of vehicles $\mathcal{K}$ (consisting of the different types). We start from an empty schedule S which is assigned a total service cost of $c = 0$. Subsequently, the requests are sorted according to the earliest possible pickup time (in ascending order). In each of the (at most) R iterations, for the request $r \in \mathcal{R}$ under consideration, we determine all feasible insertion positions $\mathcal{P}_k$ on all vehicles k with satisfied patient-vehicle compatibility constraint, i.e., $\delta_{rk} = 1$. In terms of the different insertion positions $\mathcal{P}_k$, we distinguish the following three cases where the actual delivery time T_{r_d} finally results from the actual pickup time T_{r_p} as well as the service times for loading and unloading a patient and the shortest travel times between two locations:

1. There exists no partial route for vehicle k yet, i.e., it is still located at its assigned depot. The route is then planned such that the actual pickup time is as early as possible taking into account the time windows, the announcement time and the operating time windows of the vehicle.

2. A partial route has already been planned for vehicle k, with patient r' served last. The route may be continued so that the vehicle waits at the delivery location of r' up to $x \in \{0, 5, 10, 15\}$ min before going on a direct route to ℓ_{r_p}. If more than one of these variants is feasible, the one with the earliest actual pickup time is added to $\mathcal{P}_k$. However, if none of these possibilities is feasible, a valid insertion position may result from the situation where the vehicle returns to its assigned depot, so that case 1 applies. Concerning this case, observe that a 5-min discretization of the waiting time is not mandatory, but it is considered appropriate in conformity with the subdivision of the operating day in time intervals of 5 min (cf. Sect. 4.1).
3. As before, we assume that a partial route has already been designed for vehicle k, which provides that patient r' is delivered last. The route may now be (re-) planned to allow for a joint transport of the two patients r' and r. Note, however, that for sharing a vehicle, $\lambda_{r'} = \lambda_r = 1$ must hold, as well as the resulting vehicle utilization must not exceed the capacity. Recall that (in a previous iteration) a pickup and delivery time has already been specified for r'. However, instead of performing the (already planned) delivery of r', patient r is picked up first and then both patients on board are delivered one after the other. In terms of the delivery order, each feasible option is then added to $\mathcal{P}_k$. Note that in flavor of the semi-online setting, we follow the rule that once a patient has been assigned, its pickup time remains unchanged, whereas the delivery time may be postponed. Thus, the service cost of an assignment results not only from request r, but also from r' due to its delayed delivery. A similar procedure is applied if more than two patients are to be transported together. Clearly, the stated precedence rule excludes some feasible assignments. However, relaxing this rule increases the service cost of request r' due to its late pickup. Accordingly, such an allocation is not of interest as we are ultimately interested in the best vehicle composition.

If there is no feasible insertion position for request r on any of the vehicles, we terminate and report that no feasible solution to the problem was found. Otherwise, among all feasible insertion positions $p \in \bigcup_{k \in \mathcal{K}} \mathcal{P}_k$ with minimal increase in total service cost, Δc^*, we consider all those where the vehicle has the least (additional) operating time, Δo^*. From this reduced set (which may consist of more than one element), we randomly select our final insertion position for request r. Subsequently, we update the schedule $\mathcal{S}$ as well as the total service cost c and proceed to the next iteration.

Since the approach described in Algorithm 1 does not guarantee a deterministic solution due to random decisions in (possibly) each iteration, we may apply the entire procedure multiple times to improve the quality of the solution. The best solution found across all runs is then chosen as our final solution.

Algorithm 1 Selective insertion heuristic: *best vehicle*

Input: Instance of the routing and scheduling problem, vehicle fleet $\mathcal{K}$
Output: Best solution found or `INFEASIBLE`

Initialize: $\mathcal{S} \leftarrow$ empty schedule, $c \leftarrow 0$
Sort $\mathcal{R}$ according to e_{r_p} (earliest first)
for $r = 1, \ldots, R$ **do**
    for all $k \in \mathcal{K}$ **do**
        if $\delta_{rk} = 1$ **then**
            Find all feasible insertion positions $\mathcal{P}_k$ for r on vehicle k
    $\mathcal{P} \leftarrow \bigcup_{k \in \mathcal{K}} \mathcal{P}_k$
    if $\mathcal{P}$ is empty **then**
        return `INFEASIBLE`
    else
        for all $p \in \mathcal{P}$ **do**
            Calculate $\Delta^p c$ and $\Delta^p o$
        $\Delta c^* \leftarrow \min_{p \in \mathcal{P}} \{\Delta^p c\}$
        $\Delta o^* \leftarrow \min_{p \in \mathcal{P}} \{\Delta^p o : \Delta^p c = \Delta c^*\}$
        Choose p^* randomly from all $p \in \mathcal{P}$ with $\Delta^p c = \Delta c^*$ and $\Delta^p o = \Delta o^*$
        Update $\mathcal{S}$ according to p^*
        $c \leftarrow c + \Delta c^*$
return $\mathcal{S}$ and c

The second variant of our heuristic (*best request*) differs from the first one (*best vehicle*) in that it does not loop through a sorted list of requests in order to determine the request's best insertion position. Instead, it selects in each iteration the request to be inserted next *among all* unassigned ones. In this process, a request whose best assignment to a vehicle, i.e., best insertion to a route, causes primarily the least increase in the total service cost and secondarily the least additional vehicle utilization, is selected.

We evaluated the performance of our selective insertion heuristic by comparing the best solutions attained by the two proposed variants, with the optimal solutions obtained by solving the ILP model for some rather small instances. The results confirm that the heuristic provides near-optimal objective values in a very short computation time, making it applicable to real-world instances.

4 Computational Experiments

In this section, we report on the experimental results obtained by performing our proposed approaches on a set of instances derived from real-world data. Both algorithms have been implemented in Python and run on a 64-bit Windows 10 PC with an Intel Core i5-8500T processor and 8 GB RAM memory.

Table 1 Total capacity and upper limits for each transportation mode for the different vehicle types

Capacity	$\mathcal{K}_1$	$\mathcal{K}_2$	$\mathcal{K}_3$	$\mathcal{K}_4$
Seated patients Q_k^W	3	2	3	3
Patients on a chair Q_k^C	1	1	2	0
Patients on a stretcher Q_k^S	1	1	0	0
Total capacity Q_k	3	2	3	3

4.1 Data Description

For our computational study, the Austrian Red Cross provided us with data containing information about the existing vehicle fleet as well as all transportation requests carried out in the year 2019[1] in Feldbach, a rural district in Austria.

The operating vehicle fleet consists of four different vehicle types, i.e., $\mathcal{I} = \{1, 2, 3, 4\}$: $\mathcal{K}_1$ and $\mathcal{K}_2$ represent the set of large and small ambulances, respectively, where the number of available vehicles of each type is 6, $\mathcal{K}_3$ has cardinality 3 and indicates the set of busses, and $\mathcal{K}_4$ with $|\mathcal{K}_4| = 2$ denotes the set of cars. Each of the different vehicle types has a certain total capacity as well as upper limits for each transportation mode as specified in Table 1.

We assume that all vehicles have the same operating time windows, ranging from 6 a.m. to 6 p.m. For the sake of simplicity, we discretize these time windows of 12 hours into subintervals of 5 min, which still provides sufficient accuracy.

For each request r, the provided data contains the announcement time a_r, the earliest possible pickup time e_{r_p} (which is set by the dispatcher when the transport is announced), an appointment time α_r (if available), the desired pickup and delivery locations ℓ_{r_p} and ℓ_{r_d} as well as the transportation mode q_r and the priority π_r (which is also assigned by the dispatcher). The possibility of a joint transport, the patient-vehicle compatibility constraints and the service times all depend on the patient's priority as follows: while low or medium-low priority patients can be transported jointly by all vehicle types, patients with a medium-high or high priority require an individual transport by an ambulance, i.e., a vehicle of type 1 or 2. Pickup and delivery service times are estimated from the data: for non-emergency patients, who are able to walk, the estimated pickup and delivery time is 5 min each, while for patients who need to be transported on a chair or stretcher, it is 10 min each. In contrast, for an emergency transport (regardless of the transportation mode), the estimated time required to pickup (deliver) a patient is 15 (10) min.

[1] Due to the COVID-19 pandemic, data from 2020 onward do not reflect regular operations in terms of number of requests and joint transports and are therefore not suitable for our computational experiments.

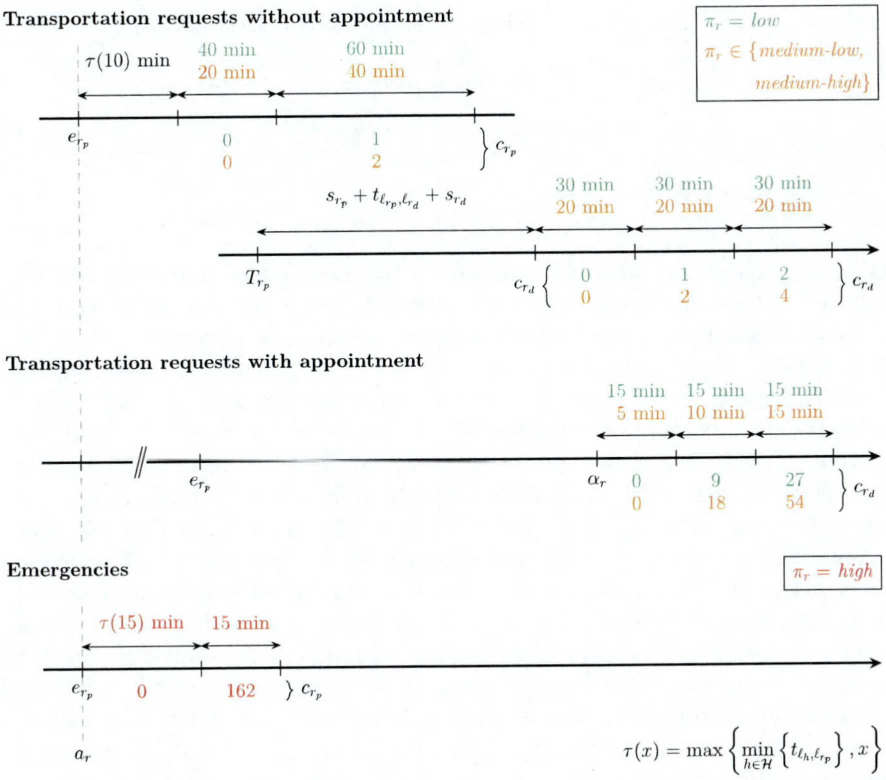

Fig. 1 Schematic illustration of how the time windows and the service costs of a single transportation request $r \in \mathcal{R}$ are specified

For each day in 2019, the travel times between all pairs of locations to be visited on that day are retrieved from *Openrouteservice*.[2] As we are only given the earliest possible pickup times, e_{r_p}, we set up the time windows in collaboration with experts from the Austrian Red Cross as described in Fig. 1. Note that we distinguish different cases depending on the stipulated priority and whether an appointment is specified or not. Beyond that, the figure expresses the service costs which depend on the actual arrival and/or departure times.

4.2 Preliminary Results

For a performance evaluation, we built 245 instances based on the days of 2019 (weekends, public holidays as well as bridging days are disregarded) where the

[2] https://openrouteservice.org.

Table 2 Comparison between the heuristic variants for the best vehicle composition

Variant	Better solution approach	Avg. infeasible solutions (over all runs)	Avg. vehicle utilization	Avg. time per run (in s)
Best vehicle	51.02 %	50.16 %	57.88 %	0.2107
Best request	48.98 %	87.08 %	57.08 %	1.8544

number of requests per day ranges from 64 to 121. For each of these instances, we evaluated a considerable number of different vehicle fleet compositions and executed 10 runs for each of them and for both heuristic variants. The computational results obtained from the best vehicle composition for every day (in terms of service quality as well as acquisition and operating costs) are reported in Table 2.

The results reveal that for 125 of the 245 instances, the best solution was achieved with the *best vehicle* variant, while the *best request* variant performed better in the remaining 120 cases. Comparing the percentage of infeasible solutions over all 10 runs on average, we find that this number is much higher for the second variant. This is, however, not surprising, as this variant selects the best request in each iteration, which conversely implies that an unselected request may not be considered until it can no longer meet the time window constraints. Moreover, an important issue in practice, especially for crew members, is the utilization of the vehicle throughout the day. As we can see from the table, however, the average vehicle utilization does not differ significantly for the two variants such that this aspect can be neglected as a criterion of quality. In contrast, there is a significant difference in the computation time which ultimately makes the main difference. Hence, we can conclude that the *best vehicle* approach is more suitable to deal with the specifications encountered in practice, both in terms of solution quality and computation time.

5 Conclusions

In this work, a routing and scheduling problem arising from the daily operations of the Austrian Red Cross has been addressed. For this purpose, the classical DARP has been extended by additional features such as a semi-online aspect, time windows of varying lengths, limited waiting times for vehicles on the route and patient-vehicle compatibility constraints. Two variants of a simple but effective insertion heuristic have been proposed that are able to cope with all characteristics of the problem. In a computational study, we evaluated the performance of the heuristic on real-world instances by comparing the best solutions attained by the two proposed variants, respectively. The results obtained from the best vehicle composition reveal that the first variant of our heuristic (*best vehicle*) is better suited for coping with the difficulties of the problem in practice, both in terms of solution quality and computation time. A possible starting point for future research might be to further develop the proposed heuristic, for instance, by incorporating it in a population-based strategy.

Beyond, future work will be devoted to develop an approach in order to predict the number of vehicles for a given day based on the requests already known the day before.

References

1. Baldacci, R., Bartolini, E., Mingozzi, A.: An exact algorithm for the pickup and delivery problem with time windows. Oper. Res. **59**(2), 414–426 (2011)
2. Cordeau, J.F.: A branch-and-cut algorithm for the dial-a-ride problem. Oper. Res. **54**(3), 573–586 (2006)
3. Cordeau, J.F., Laporte, G.: The dial-a-ride problem: models and algorithms. Ann. Oper. Res. **153**, 29–46 (2007)
4. Cosmi, M.: Scheduling for the Last-Mile Food Delivery Problem. Ph.D. thesis, Roma Tre University (2021)
5. Gschwind, T., Irnich, S.: Effective handling of dynamic time windows and its application to solving the dial-a-ride problem. Transp. Sci. **49**(2), 335–354 (2015)
6. Ho, S.C., Szeto, W.Y., Kuo, Y.H., Leung, J.M.Y., Petering, M., Tou, T.W.H.: A survey of dial-a-ride problems: Literature review and recent developments. Transp. Res. Part B: Methodol. **111**, 395–421 (2018)
7. Molenbruch, Y., Braekers, K., Caris, A.: Typology and literature review for dial-a-ride problems. Ann. Oper. Res. **259**, 295–325 (2017)
8. Parragh, S.N.: Introducing heterogeneous users and vehicles into models and algorithms for the dial-a-ride problem. Transp. Res. Part C: Emerg. Technol. **19**(5), 912–930 (2011)

Towards a Unified Framework for Routing and Scheduling Planning in an Integrated Continuous Care Unit

Maria Teresa Godinho and Maria João Lopes

Abstract We study a real world routing and scheduling problem arising in home health care context. Requisites defined by stakeholders are introduced and examined. Then, the problem is modeled as a Time-constrained Vehicle Routing Problem with Time Windows and a new Single Commodity Flow Formulation is proposed. Valid Inequalities to enhance the model are presented. Extensions of the problem, such as aggregating different types of care and of ensuring a threshold on the difference between the duration of the routes, are also discussed. We discuss the results obtained from a set of computational tests on instances adapted from the literature.

Keywords Home health care · Routing and scheduling · Mixed integer linear formulations

1 Introduction and Literature Review

The generic term "Home health care" covers several programs of care delivered at patient's home. In this work, we address a real case study motivated by an Integrated Continuous Care (ICC) Unit routing and scheduling needs. Next, we review some contributions on the routing and scheduling problem in health care context. Our review is based on [3, 6, 7].

The home health care routing and scheduling problem (HHCRSP) consists of designing a set of routes that allow to provide planned care visits to a set of patients

M. T. Godinho (✉)
Polytechnic Institute of Beja, School of Technology and Management, Rua Pedro Soares, 7800-295 Beja, Portugal
e-mail: mtgodinho@ipbeja.pt

M. J. Lopes
Iscte-IUL University Institute of Lisbon, Business School, Avenida das Forças Armadas, 1649-026 Lisboa, Portugal
e-mail: maria.joao.lopes@iscte-iul.pt

M. T. Godinho · M. J. Lopes
CMAFcIO, University of Lisbon, Campo Grande, 1749-016 Lisboa, Portugal

R. Aringhieri et al. (eds.), *Operations Research for Health Care in Red Zone*, AIRO Springer Series 10, https://doi.org/10.1007/978-3-031-38537-7_7

over a planning horizon, while optimizing some given criteria and respecting several constraints [3]. Thus, the HHCRSP can be seen as a VRP where side-constraints are used for modeling the different requirements of each application, including scheduling requirements.

Side constraints can be grouped into (i) patient constraints; (ii) caregivers constraints and (iii) treatment/visit constraints. Constraints type (i) deal with patient's preferences regarding both the visit's time-span and the caregiver; constraints of type (ii) tackle labour legislation and also the preferences of the caregivers; finally, (iii) gathers constraints stemming from specifications of the care and constraints that ensure compliance with temporal dependencies between different visits to the same patient, as each patient may require more than one visit per day. The objective functions also vary. The most frequent objectives are the minimization of the routes and/or of the staff related total cost. Other objectives present in the literature are the optimization of the quality of service and well-being at work. These objectives are handled mostly under multi-objective approaches, either by considering weighted sums or by approximating the Pareto Frontier and analysing trade-offs.

Both exact and heuristic methods can be found in the literature. In our work, we present a new Mixed Integer Linear Programming model of this problem. As mentioned, the HHCRSP can be seen as a VRP, to which side constraints have been added. Thus, many MILP models of it, in the literature, are based on models originally proposed for the VRP or its variants. This is the case of [1, 2, 9]. While Gomes and Ramos [9] present a model adapted from a widely used formulation of the VRP with Time Windows proposed in [5], Cappanera and Scutellà and Cappanera et al [1, 2] present flow based models rooted in the work of Wong [12] and Claus [4] and of Gavish and Graves [8], respectively.

Pure MILP approaches are seldom seen in recent papers. The references cited above, with the exception of [2], use the proposed models combined with other approaches. For fairness, it should be said that the conclusions of [2] point in the same direction. Nevertheless, in this work, our model is solved by feeding it to CPLEX. In fact, we use the proposed model as an exploratory mean to gain a better understanding as to the many aspects of the problem, but we are fully aware that to solve medium size instances in reasonable times we need to go a step further.

The remainder of this work is structured as follows: in Sect. 2, the case study that motivated this work is described in detail; in Sect. 3, the problem is formally defined, a MILP formulation is introduced and its extensions discussed; in Sect. 4, a computational study is presented and Sect. 5 concludes the text.

2 Case Study

As mentioned, this work tackles a real case study originated in a ICC Unit. In order to understand the requisites of the problem, the Head of the Transport Department, the Head Nurse and the drivers were heard. Due to the existence of many conflicting aspects, it was decided that the Head Nurse requests should be given priority.

The Head Nurse pointed out that some patients have to be visited more than once on the same day and that while some treatments can be performed at any given time, others have to be carried out within a specific time-slot. She also indicated that the number of teams is fixed, but the composition of the teams may vary. The allocation of the personnel to the teams is done in compliance with their skills, ensuring that the teams are balanced, and also with work shifts and days off. As it is, any team can be assigned to any patient. Enquired about the importance of continuity of care (we say that there is when the same patient is treated by the same team), the Head Nurse said that all ICC members know quite well all patients, because ICC teams deal with long term care and are composed of dedicated personnel. Thus, for the time being, she does not feel the need to introduce that type of constraints. The teams are out for the maximum duration of the shift and use a vehicle with a driver to travel to and between patients' homes. Finally, she elected as her objective for this project, the desire to increase the percentage of time actually dedicated to treating the patients.

In conclusion, requisites are: minimization of wasted time, while meeting the time windows requirements and the bounds on both the duration of the routes and the number of teams.

3 Modeling Approach

The problem described in the previous section can be modeled as a Time-constrained Vehicle Routing Problem with Time-windows (`TCVRPTW`). The `TCVRPTW` seeks a set of routes of minimum cost in a given network, such as requirements on time-windows, number of vehicles and duration of the routes are met.

Recall that any team can carry out any treatment to any patient but there are treatments with and without time windows associated with them. Thus, we consider two types of treatments: those with time windows, designated by specific (s), and the others, designated by general (g).

Consider the set $P = \{1, ..., p\}$ of the patients to treat on a given day and the set $C = \{s, g\}$ of the types of care the unit can provide. Thus, p^c, $p \in P$, $c \in C$, define the tasks to be executed by the team. As patients may need more than one care on the same day, there may be more than one task associated with the same patient.

Let $G = (N, A)$ be a complete directed graph with $N = \{0, ..., n\}$. Node 0 represents the depot and the remaining nodes represent the tasks. We let N^s and N^g denote the subsets of the nodes associated with tasks type s and g, respectively. d_i indicates the time required to complete task $i \in N \backslash \{0\}$ and $[a_i, b_i]$ denotes the time window within which the task $i \in N^s$ has to be concluded. The arcs represent the trips. t_{ij} indicates the traveling time between the homes of the patients associated with tasks i and j, for all $(i, j) \in A$. When tasks i and j are associated with the same patient, t_{ij} is null. Finally, T denotes the maximum duration of a route, which is given by the duration of the shift. The number of teams defines the size of the fleet, denoted by V. We assume that all parameters are non-negative integers (the time unit is minutes).

Consider, also, the decision variables:

- binary variables, $x_{ij} = 1$ if j is visited just after i; $x_{ij} = 0$, otherwise, for all $i, j \in N, i \neq j$;
- variables y_{ij} representing the arrival time at j just after visiting i, for all $i, j \in N, i \neq j$;
- w_{ij} representing the waiting time at j, after traveling from i to j, for all $i \in N$, $j \in N^s, i \neq j$.
- binary variables $u_{p^s,p'^g} = 1$, if care s is provided before care g to a patient requiring both types of care, $u_{p^s,p'^g} = 0$, otherwise, for all $p^s \in N^s$, $p'^g \in N^g$, with $p = p'$.

Then, the feasibility set of TCVRPTW can then be modeled as:

$$\sum_{i\in N} x_{ij} = 1 \qquad j \in N\backslash\{0\} \tag{1}$$

$$\sum_{j\in N} x_{ij} = 1 \qquad i \in N\backslash\{0\} \tag{2}$$

$$y_{0j} = t_{0j}x_{0j} \qquad j \in N\backslash\{0\} \tag{3}$$

$$\sum_{i\in N}(y_{ij} + w_{ij} + d_j x_{ij}) = \sum_{i\in N}(y_{ji} - t_{ji}x_{ji}) \qquad j \in N^s \tag{4}$$

$$\sum_{i\in N}(y_{ij} + d_j x_{ij}) = \sum_{i\in N}(y_{ji} - t_{ji}x_{ji}) \qquad j \in N^g \tag{5}$$

$$y_{j0} \leq T x_{j0} \qquad j \in N\backslash\{0\} \tag{6}$$

$$y_{ij} + w_{ij} \leq (T - d_j - min_l\{t_{jl}\})x_{ij} \qquad i \in N, j \in N^s, i \neq j \tag{7}$$

$$y_{ij} \leq (T - d_j - min_l\{t_{jl}\})x_{ij} \qquad i \in N, j \in N^g, i \neq j \tag{8}$$

$$\sum_{i\in N}(y_{ij} + w_{ij}) \leq b_j - d_j \qquad j \in N^s \tag{9}$$

$$\sum_{i\in N}(y_{ij} + w_{ij}) \geq a_j \qquad j \in N^s \tag{10}$$

$$\sum_{i\in N}(y_{ik^s} + w_{ik^s}) + d_k \leq \sum_{i\in N}(y_{il^g}) + M(1 - u_{k^s l^g}) \qquad k^s \in N^s, l^g \in N^g, k = l \tag{11}$$

$$\sum_{i\in N} y_{il^g} + d_l \leq \sum_{i\in N} y_{ik^s} + M u_{k^s l^g} \qquad k^s \in N^s, l^g \in N^g, k = l \tag{12}$$

$$\sum_{j\in N\backslash\{0\}} x_{0j} \leq V \tag{13}$$

$$x_{ij} \in \{0, 1\}, y_{ij} \geq 0 \qquad i, j \in N, i \neq j \tag{14}$$

$$w_{ij} \geq 0 \qquad i \in N, j \in N^s, i \neq j \tag{15}$$

$$u_{i^s, j^g} \in \{0, 1\} \qquad i^s \in N^s, j^g \in N^g, i = j \tag{16}$$

Time is modeled as a flow system whose source and sink are node 0. The flow values increase as patients are visited and cared. Assignment constraints, (1) and (2), ensure that all tasks are executed; flow conservation constraints, (3), (4) and (5), ensure the connectivity of the routes and, thus, time continuity; linking constraints, (6), (7) and (8) ensure that the duration of a route does not exceed T and, together with assignment and flow conservation constraints, prevent the presence of sub-circuits; constraints (9) and (10) ensure that the time windows are satisfied and constraints (11) and (12) prevent that patients are treated for two types of care at the same time; constraints (13) ensure that the total number of vehicles is not exceeded and, finally, constraints (14), (15) and (16) define the variables domain.

As mentioned, we aim at minimizing "wasted" time. With that purpose, we consider two criteria: minimizing the total traveling time, (17), and minimizing the total waiting time, (18) that will be studied separately.

$$\min F_1 = \sum_{i,j \in N, i \neq j} t_{ij} x_{ij} \tag{17}$$

$$\min F_2 = \sum_{i \in N, j \in N^s, i \neq j} w_{ij} \tag{18}$$

The corresponding ILP models are:

$$\min F_i$$
$$\text{subject to} \quad (1) - (16)$$

with $i = 1, 2$.

3.1 Enhancing the Models

The linear programming relaxation of the models described above can be enhanced by adding valid inequalities to it. Consider the following set of inequalities:

$$(min_l\{t_{lj}\} + t_{ji} + d_j) x_{ji} \leq y_{ji} \qquad j \in N \backslash \{0\}, i \in N, i \neq j \tag{19}$$

Constraints (19) model the fact that if node i is not the first to be visited (that is if $x_{ji} = 1$ to some $j \neq 0$), then some time has already been consumed in the previous travels and treatments. That time cannot be inferior to $min_l\{t_{lj}\} + t_{ji} + d_j$ for $l \neq i, j$. Similarly, observing that the care associated with nodes in N^s cannot start outside their time window, we have:

$$(a_j + t_{ji} + d_j)x_{ji} \leq y_{ji} \qquad j \in N^s, i \in N, i \neq j \tag{20}$$

Thus for the nodes in N^s constraints (19) can be replaced by

$$[max(min_l\{t_{lj}\}, a_j) + t_{ji} + d_j]x_{ji} \leq y_{ji} \qquad j \in N^s, i \in N, i \neq j \tag{21}$$

Another set of valid inequalities is:

$$y_{ji} \leq (b_j + t_{ji})x_{ji} \qquad j \in N^s, i \in N, i \neq j \tag{22}$$

Preliminary tests, showed that, in most cases, the improvements on the lower bounds and computational times associated with adding (19), (21) and (22) to the models are substantial. Thus, in the following, we consider the enhanced version of the two formulations.

3.2 Model Extensions

We have considered two different model extensions:

- Aggregating general and specific treatments, when applied to the same patient;
- Ensuring time work balance within the teams.

Firstly, we observe that any solution, of one of the models, in which a general care and a specific care associated with the same patient are carried out in the same visit, can never be more costly than a solution where the two cares are carried out in different visits. That is, the latter will occur only when the former is unfeasible. Thus, a new version of the model, where the two cares are aggregated was written, by making just a few adjustments to the current formulations. As a result of the decrease of the size of the network and of the elimination of constraints (11) and (12) the time consumed to solve the model to optimality was reduced considerably. Consequently, the computational experiments in Sect. 4 were performed using the new version of the formulations as the base model. If unfeasibility would occur in a given instance (which did not happened) then we would resort to the previous model to solve the instance.

Secondly, large differences in the duration of the routes were observed in some of the solutions obtained in the preliminary tests. In consequence, a new set of constraints imposing a threshold on the difference between the duration of the routes, was added to the model. We considered a threshold of 30 min:

$$D_{min} \leq y_{j0} + M(1 - x_{j0}) \qquad j \in N\backslash\{0\} \tag{23}$$

$$y_{j0} \leq D_{max} \qquad j \in N\backslash\{0\} \tag{24}$$

$$D_{max} - D_{min} \leq 30 \tag{25}$$

4 Computational Study

The models presented in the previous sections were analyzed empirically with the goal of characterizing the solutions produced in each case and the associated trade-offs. Experiments were carried out on a 64-bit PC with an Intel®Core™ i7-7700HK Quad core at 2.8GHz with 16GB of RAM. We have used IBM ILOG CPLEX version 22.1 as the ILP solver, with no time limit. Real instances are not available due to confidentiality issues, regarding patient's data. Thus, the instances considered were adapted from the literature. We have used the first six instances of the C2 set of the Solomon instances [11] and performed the following adjustments:

1. The size of the instances was reduced to 45:30 of them require specific care only; 10 require general care only and 5 require both. Patients in each category were randomly chosen among the first 45 individuals of the original instances;
2. The values of the traveling times were the rounded to the next integer;
3. The total number of vehicles was set to 13;
4. The maximum duration of a route was set to 540 min;
5. The duration of cares was a random integer between 30 and 60 min;
6. The time-window of each patient $i \in N^s$ was generated by: (i) obtaining the center of the interval, generating a random integer between t_{0i} and $540 - t_{i0}$; (ii) obtaining the width of the interval, generating a random integer between 60 and 120; (iii) correcting unfeasibilities: if a_i is negative, $-a_i$ is added to both bounds; if $t_{0i} + d_i < b_i$, b_i is perturbed until unfeasibility ceases to exist.

The values introduced in adjustments 4 and 5 were supplied by the stakeholders; the procedure described in 6 is suggested in [10] for a similar problem (in the original set of instances, all time windows have the same amplitude, which is not realistic); the value proposed in 1 is arbitrary; the value proposed in 3 was changed accordingly. Results for this set of tests and the corresponding computing times (in seconds) are reported in Tables 1, 2, 3, 4, 5 and 6. In all these tables, MnRD, AvRD and MxRD denote the minimum, the average and the maximum routes duration, TTT and TWT the total traveling and waiting times, respectively, and NR the number of routes.

In Table 1 we present the results obtained when the total traveling time was minimized. The number of routes was 8 for two instances and 9 for the remaining four. The routes duration ranged from 167 to 540 min. In addition, the maximum duration for all instances was greater than 500 min. The total waiting time, for each instance, was greater than the total traveling time. On the other hand, to meet time windows there were several routes with waiting time before the first care, because each route started at 0. This means that several routes can start after 0 and their duration should not include the previous waiting times. All instances were solved in less than two minutes.

Table 1 Results for the minimization of the total traveling time

Instances	C201-45	C202-45	C203-45	C204-45	C205-45	C206-45
MnRD	426	236	429	173	167	255
AvRD	488.1	407.7	477.3	409.9	429.3	451.9
MxRD	540	540	502	540	540	535
TTT	685	681	649	699	716	718
TWT	1012	736	939	801	762	1068
NR	8	9	8	9	9	9
Time	69	104	30	29	40	66

Table 2 Results for the minimization of total traveling time excluding the waiting time before the first care

Instances	C201-45	C202-45	C203-45	C204-45	C205-45	C206-45
MnRD	309	76	309	171	71	89
AvRD	439.6	360.9	445.4	388	396.2	421.3
MxRD	540	540	498	528	506	535
TTT	685	681	649	699	716	718
TWT	624	315	684	604	464	793
NR	8	9	8	9	9	9

Table 3 Results for the minimization of total traveling time with routes duration balance

Instances	C201-45	C202-45	C203-45	C204-45	C205-45	C206-45
MnRD	476	464	472	467	471	496
AvRD	483.6	481.7	485.4	476.2	498.9	510.8
MxRD	506	491	502	497	501	526
TTT	693	719	656	743	726	737
TWT	968	1364	997	1354	807	1068
NR	8	9	8	9	8	8
Time	99	1406	49	47	91	89

Table 4 Results for the minimization of total traveling time with routes duration balance and excluding the waiting time before the first care

Instances	C201-45	C202-45	C203-45	C204-45	C205-45	C206-45
MnRD	246	294	359	387	416	441
AvRD	405.4	398.4	457.3	466.0	471.0	494.0
MxRD	484	489	501	497	501	526
TTT	693	719	656	743	726	737
TWT	342	615	772	1262	656	934
NR	8	9	8	9	8	8

Table 5 Results for the minimization of the total waiting time with routes duration balance

Instances	C201-45	C202-45	C203-45	C204-45	C205-45	C206-45
MnRD	463	469	464	458	467	481
AvRD	481.3	484.9	476.6	469.1	480.1	492.9
MxRD	493	499	493	488	488	504
TTT	1642	1613	1583	1552	1455	1659
TWT	0	14	0	12	0	3
NR	8	8	8	8	8	8
Time	955	1561	16329	1598	16915	>63491

Table 6 Results for the minimization of total traveling time, constraining the total waiting time to the minimum, with routes duration balance, for the first 5 instances

Instances	C201-45	C202-45	C203-45	C204-45	C205-45
MnRD	470	464	466	458	467
AvRD	480.4	475.0	476.9	463.3	475.3
MxRD	491	491	488	497	497
TTT	1155	1059	1108	1506	941
TWT	0	14	0	12	0
NR	7	7	7	8	7
Time	>18609	>53030	>27822	>173295	96725

In Table 2 we present the results obtained excluding the waiting times before the first care. The waiting times are smaller but they are still significant. This correction led to routes duration ranging from 71 to 540 min. So, the exclusion of initial waiting times increased the differences in the duration of the routes.

In Table 3 we present the results when the total traveling time was minimized and a gap for routes duration of at most 30 min was imposed. For all instances but one the maximal gap was attained. The exception was a gap equal to 27 min. The number of instances with 8 routes increased to 4. The others kept the 9 routes. For each instance, the total waiting time was greater than the total traveling time and several routes had waiting time before the first care. Five out of the six instances were solved in less than two minutes. However, the most time consuming instance took nearly 24 min to be solved to optimality.

In Table 4 we present the results excluding the waiting times before the first care. In spite of the reduction on the total waiting time, for all instances, these times are still very high. In addition, the correction on routes duration showed that their gap are greater than 30 min. To try to overcome this disadvantage, the objective was changed to the minimization of the total waiting time.

Table 5 reports these results. The total waiting time reduced significantly. For three instances, in one of the routes there was a waiting time before the first care, but these times were at most equal to 12 min. These times had a little effect on routes duration gaps. The number of routes was equal to 8. Thus, there was a reduction for 2 instances. However, the total traveling time increased significantly, it exceeded twice the total traveling time presented in Tables 1 and 3. On the other hand, for all instances but one the total traveling time reported in Table 5 was lower than the corresponding total time (traveling plus waiting times) represented in Table 3. The computing times increased significantly, they range from 15 min to more than 17 h. Minimizing the total waiting time seems to be more difficult than minimizing the total traveling time.

Notice that the computational results analyzed so far minimize one objective (traveling or waiting time) without constraining other objectives. To keep the advantages of the previous solutions, preliminary computational experiments were carried out. The objective was to minimize the total traveling time, constraining the total waiting time to the minimum, and avoiding significant differences in the duration of the routes. The results obtained so far for the first 5 five instances are presented in Table 6. Notice that for the four first instances, we present data concerning the best feasible solution obtained. In general, the reduction on the total traveling time was very significant. Moreover, for four instances, the number of routes decreased to 7. The computing times reported in this table are much greater than the ones present in the previous table. Minimizing the total traveling time constrained to the minimum total waiting time seems to be even more complex than minimizing the total waiting time.

5 Conclusions and Future Work

We addressed the routing and scheduling problem of an ICC Unit, by means of a Single Commodity Flow Formulation (*SCF*). SCF models have the advantage of being simple to adapt to the many extensions of this problem and easy to solve.

Results permitted to analyze trade-offs as conditions varied. Results are now to be analyzed with the Head Nurse and the remaining stakeholders in order to understand what still needs to be accommodated in the model. No changes are to be made definitive without the Head Nurse approval.

In the near future, we plan to generate more and more realistic instances. Namely, it is necessary to study the effect on the results of varying the percentages of each type of patient in the data. In the long range, we intend to introduce uncertainty in the model and to develop a hybrid approach that combines MILP with heuristic methods.

References

1. Cappanera, P., Scutellà, M.G., Nervi, F., Galli, L.: Demand uncertainty in robust home care optimization. Omega **80**, 95–110 (2018). https://doi.org/10.1016/j.omega.2017.08.012, www.sciencedirect.com/science/article/pii/S0305048316309008
2. Cappanera, P., Scutellà, M.G.: Joint assignment, scheduling, and routing models to home care optimization: A pattern-based approach. Transp. Sci. **49**, 830–852 (2015). https://doi.org/10.1287/trsc.2014.0548, https://pubsonline.informs.org/doi/epdf/10.1287/trsc.2014.0548

3. Cissé, M., Yalçındağ, S., Kergosien, Y., Şahin, E., Lenté, C., Matta, A.: Or problems related to home helth care: A review of relevant routing and scheduling problems. Oper. Res. Health Care **13–14**, 1–22 (2017). https://doi.org/10.1016/j.orhc.2017.06.001, www.sciencedirect.com/science/article/pii/S2211692316300157
4. Claus, A.: A new formulation for the travelling salesman problem. SIAM J. Algebr. Discret. Methods **5**(1), 21–25 (1984). https://doi.org/10.1137/0605004
5. Cordeau, J.F., Desaulniers, G., Desrosiers, J., Solomon, M.M., Soumis, F.: VRP with time windows. In: SIAM Monographs on Discrete Mathematics and Applications, pp. 157–193, Philadelphia (2002). https://doi.org/10.1137/1.9780898718515.ch7, https://epubs.siam.org/doi/abs/10.1137/1.9780898718515.ch7
6. Di Mascolo, M., Martinez, C., Espinouse, M.L.: Routing and scheduling in home health care: A literature survey and bibliometric analysis. Comput. Ind. Eng. **158**, 107255 (2021). www.sciencedirect.com/science/article/pii/S0360835221001595
7. Fikar, C., Hirsch, P.: Home health care routing and scheduling: a review. Comput. Oper. Res. **77**, 86–95 (2017). https://doi.org/10.1016/j.cor.2016.07.019, www.sciencedirect.com/science/article/pii/S0305054816301848
8. Gavish, B., Graves, S.C.: The travelling salesman problem and related problems. Technical Report, Massachusetts Institute of Technology (1978). http://hdl.handle.net/1721.1/5363
9. Gomes, M.I., Ramos, T.R.P.: Modelling and (re-)planning periodic home social care services with loyalty and non-loyalty features. Eur. J. Oper. Res. **277**(1), 284–299 (2019). https://doi.org/10.1016/j.ejor.2019.01.061, www.sciencedirect.com/science/article/pii/S0377221719300967
10. Li, Y., Xiang, T., Szeto, W.Y.: Home health care routing and scheduling problem with the consideration of outpatient services. Transp. Res. Part E: Logist. Transp. Rev. **152**, 102420 (2021). https://doi.org/10.1016/j.tre.2021.102420, www.sciencedirect.com/science/article/pii/S1366554521001861
11. Solomon, M.M.: Algorithms for the vehicle routing and scheduling problems with time window constraints. Oper. Res. **35**, 254–265 (1987)
12. Wong, R.T.: Integer programming formulation of the travelling salesman problem. In: Proceedings IEEE International Conference of Circuit and Computers, pp. 149–152 (1980)